W9-BLH-907

WHY DOES ASPARAGUS MAKE YOUR WEE SMELL?

WHY DOES ASPARAGUS MAKE YOUR WEE SMELL?

And 57 other curious food and drink questions

ANDY BRUNNING

To Anna, for her continued support and
enthusiasm throughout the creation of this book.

CONTENTS

MIND

HEALTH

TRANSFORMATION

INTRODUCTION

Food plays a huge part in our everyday lives, but we seldom think of the science behind it. We all know that chopping onions makes you cry, that eating garlic gives you bad breath and that mint makes your mouth feel cool, but many would struggle to describe the chemical reasons behind these strange effects. The aim of this book is to look at the quirky and sometimes downright weird properties that food and drink can exhibit, and explain in simple terms the chemistry that leads to them.

In our explanations, we're going to be talking primarily about organic chemistry. This use of the word 'organic' is different from the one that we usually associate with food in supermarkets; in chemistry, it refers to chemical compounds based on carbon. There are many millions of possible organic compounds, with a wide range of different properties. You, yourself, are made up of organic compounds, and so are the foods we eat. These compounds are responsible for the flavour and aroma of the foods and drinks we consume on a daily basis, and can also help explain some of the different effects that we see.

The hope is that the material in this book will be accessible even to those who only have a basic understanding of chemistry. There is a brief introduction on the next page to some of the chemistry, to help with interpretation of the structures and descriptions later in the book.

If you do have a background in chemistry, and want to read in further detail on some of the topics examined here, references to the studies consulted during the writing of this book are also given at the end.

Whatever your interests or curiosities, I hope that the chemistry in this book will turn the everyday into the extraordinary for you.

A BRIEF INTRODUCTION TO ORGANIC CHEMISTRY

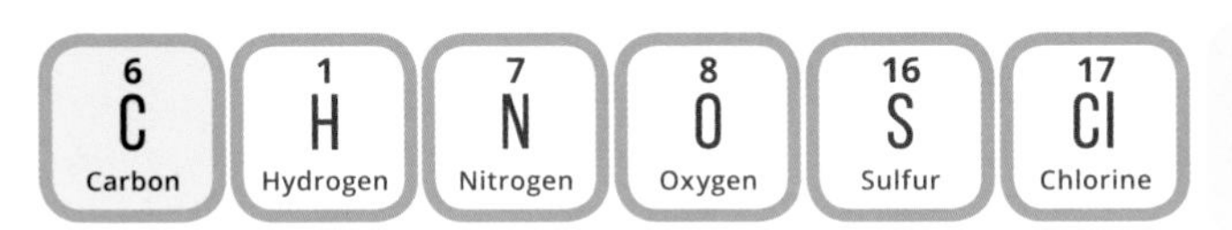

Organic chemistry is the study of carbon-based compounds. Organic compounds usually contain primarily carbon and hydrogen, but we'll also see some containing the other elements shown here.

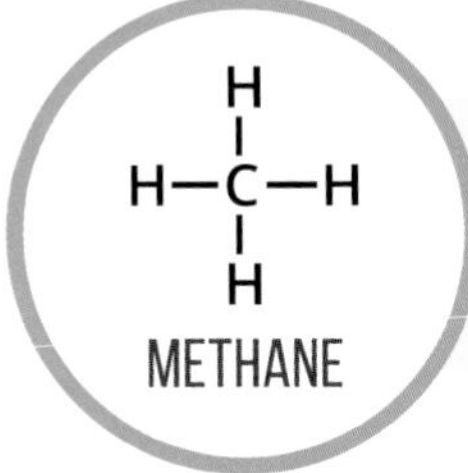

The chemical bonds in organic compounds are formed by the atoms sharing a pair of electrons with each other. Carbon can form four bonds to other atoms; oxygen usually forms two bonds, while hydrogen can only form one bond. Bonds are shown as lines; two lines between atoms indicates a double bond, or two shared electron pairs.

Displayed formula

Skeletal formula

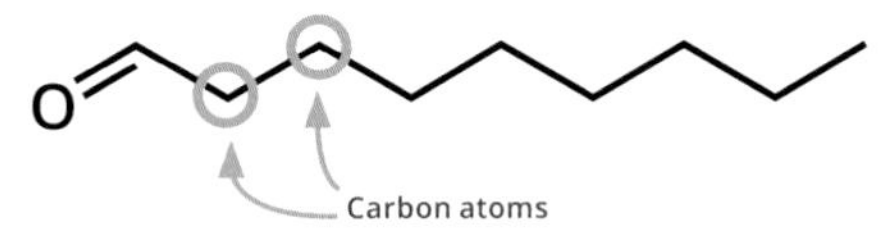

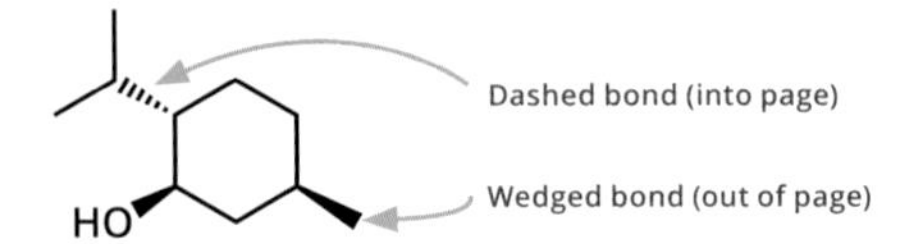

Organic compounds can be drawn showing all their bonds and all their atoms, as on the left. However, for bigger molecules, this can end up looking quite complicated so we commonly use skeletal formula to represent molecules.

Skeletal formula show each carbon atom as a bend in the chain of the molecule. Hydrogen atoms attached to carbons aren't shown, to make the structure easier to interpret. Atoms other than carbon and hydrogen are shown.

While, for ease, chemical structures are shown in two dimensions on the page, obviously they are three dimensional in real life. In some cases, it can be useful to indicate this in structures, so for this purpose we can sometimes use dashed bonds, to show a bond going away from you and into the page, and wedged bonds, to show a bond coming towards you and out of the page.

FUNCTIONAL GROUPS IN ORGANIC CHEMISTRY

Functional groups are groups of atoms in organic compounds that are responsible for the characteristic reactions and properties of those compounds. In many compounds in this book, you will see several of these groups in one molecule. The functional groups present in a molecule are usually reflected in its name, as indicated here.

You may also see 'R' used in some molecules – this represents a further part of the molecule that may be variable. If you see 'X' used, it indicates a halogen atom (the halogens are the elements fluorine, chlorine, bromine and iodine). Some typical functional groups are shown here.

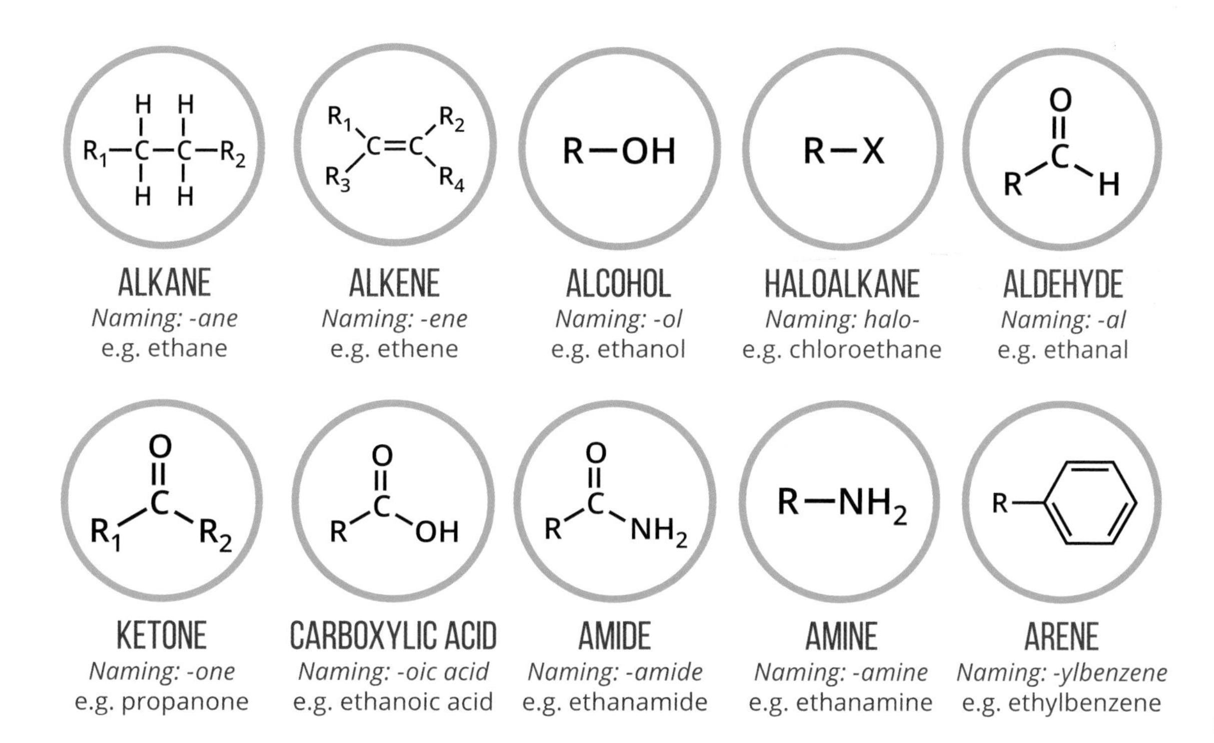

FLAVOUR

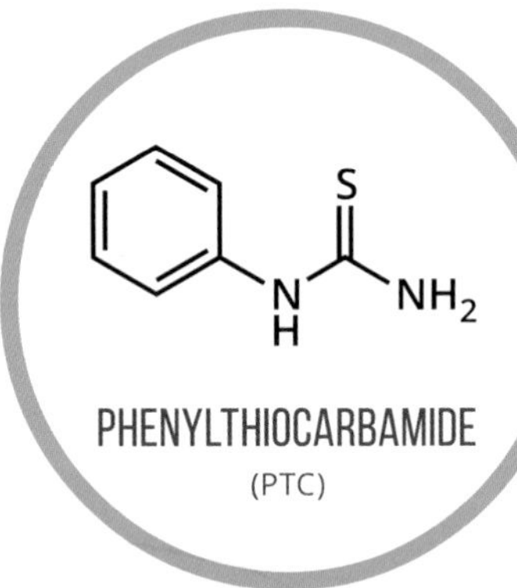

PHENYLTHIOCARBAMIDE

(PTC)

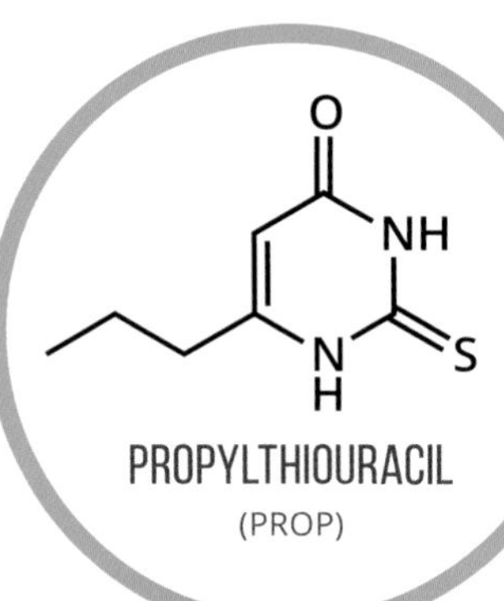

PROPYLTHIOURACIL

(PROP)

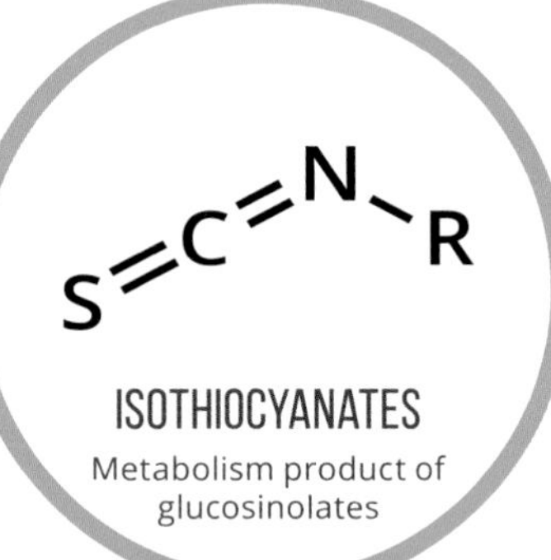

ISOTHIOCYANATES

Metabolism product of glucosinolates

PTC TASTES BITTER TO 70% OF PEOPLE

It's thought that cruciferous vegetables can induce a similar effect, as their breakdown products, isothiocyanates, are chemically similar.

EXAMPLES OF CRUCIFEROUS VEGETABLES

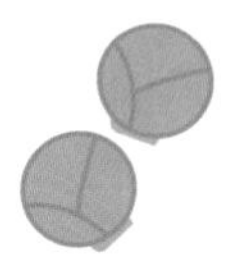

SPROUTS

BROCCOLI

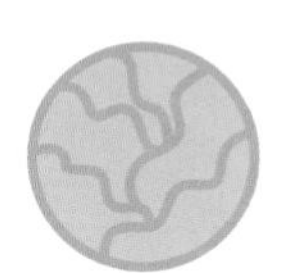

CABBAGE

CAULIFLOWER

Brussels sprouts generally have higher levels of glucosinolates compared to most other cruciferous vegetables

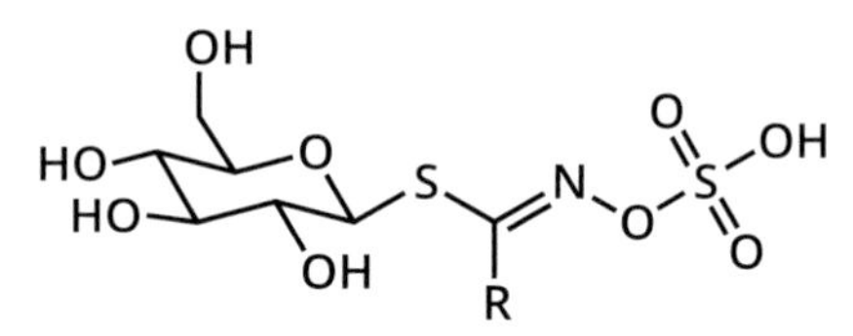

GLUCOSINOLATES

Found in all cruciferous vegetables (R group varies)

WHY DO SOME PEOPLE HATE BRUSSELS SPROUTS?

There's one vegetable at the Christmas dinner table that's always bound to elicit strong and contrary opinions: Brussels sprouts. Much like Marmite, they seem to conjure up a 'love it or hate it' sentiment; however, if you fall into the latter camp there may be a chemical and genetic reason why you can't stand the taste.

Before discussing Brussels sprouts specifically, we actually need to look at a chemical that isn't even found in them, phenylthiocarbamide (PTC). This compound is an oddity in that it tastes bitter – but only to around 70 per cent of people. To the other 30 per cent it's completely tasteless. This property of PTC was discovered by accident in 1931, when Arthur Fox, a chemist working for the chemical company DuPont, accidentally spilt some of the compound while he was working with it. His colleague working nearby complained of the bitter taste, but Fox couldn't detect it.

Fox went on to carry a series of tasting tests with his friends and family, further confirming that it tasted bitter to some, but not all. These studies, along with further work, confirmed the ability to taste PTC as a dominant genetic trait, and one that can be inherited. Before the advent of readily available DNA testing, the ability to taste PTC was commonly used in paternity cases. Another compound, propylthiouracil (PROP), is similar in that it also tastes bitter to some and not others, and is now the more commonly used compound in taste research.

At this point, you're probably wondering what this has to do with Brussels sprouts. While PTC and PROP aren't found in vegetables, they contain a thiocyanate group (nitrogen, carbon and sulfur bonded in series) which is thought to be related to its bitter taste. This same group is also present in compounds called glucosinolates, which occur naturally in Brussels sprouts, as well as broccoli, cabbage and kale (collectively known as cruciferous vegetables). There also seems to be a strong association between the ability to detect the bitter taste of PROP and a sensitivity to the bitterness of these vegetables.

It's not the case that sensitivity to PTC and PROP automatically implies a dislike of cruciferous vegetables; other environmental factors may also have an effect. However, if you're a Brussels sprout hater and you are berated for your perceived vegetable prejudice, you can now suggest a chemical reason for your dislike.

WHY DO ARTICHOKES MAKE DRINKS TASTE SWEETER?

Artichokes have an unusual claim to fame – they're the only known plants consumed on a large scale across the world that have taste-modifying properties. To some people it can cause drinks to taste sweeter straight after eating it, which apparently causes great difficulty when pairing dishes containing artichokes with wine. This effect is down to particular chemical constituents contained within artichokes.

This strange effect was first noted in research at an American Association for the Advancement of Science dinner. Artichokes were included in one of the courses, and after it was said that 60 per cent of the 250 people attending reported that the water being served tasted sweeter. Decades later, a researcher investigated this effect by exposing subjects to artichoke extract, as well as individual constituents of the extract, and recording the perceived impact on the taste of water.

It was found that, of the constituent compounds in artichokes, potassium salts of chlorogenic acids and cynarin were major causes of the effect. Potassium ions are the dominant metal ions present in artichokes, hence why potassium salts of the compounds were used. Cynarin in particular was found to have a sweetening effect equivalent to adding two teaspoons of sugar to 170 millilitres of water. That said, although these compounds were found to be the major sources, they still didn't account for the full effect of artichoke, so other, lesser compounds may also be implicated.

We still don't know exactly how these compounds cause other substances to taste sweeter, but we do know it's a result of them interacting with sweet-detecting taste buds on the tongue in such a way that non-sweet substances taste sweet. It also doesn't affect everyone, as the survey taken at the dinner which initially highlighted the effect showed – it's assumed that there must be some form of genetic basis to the effect.

So, the next time you get a chance, why not try eating an artichoke and see if it works for you?

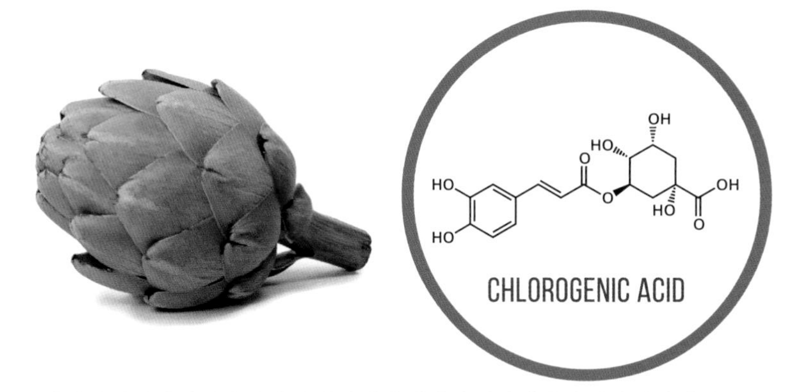

APPROX. 60%

Estimated percentage of the population who experience the sweetening effect of artichokes

CYNARIN

Major compound that affects sweetness

THE COMPOUNDS IN ARTICHOKES HAVE A SIMILAR SWEETENING EFFECT ON TASTE BUDS TO ADDING **TWO TEASPOONS OF SUGAR** TO **170ML OF WATER**

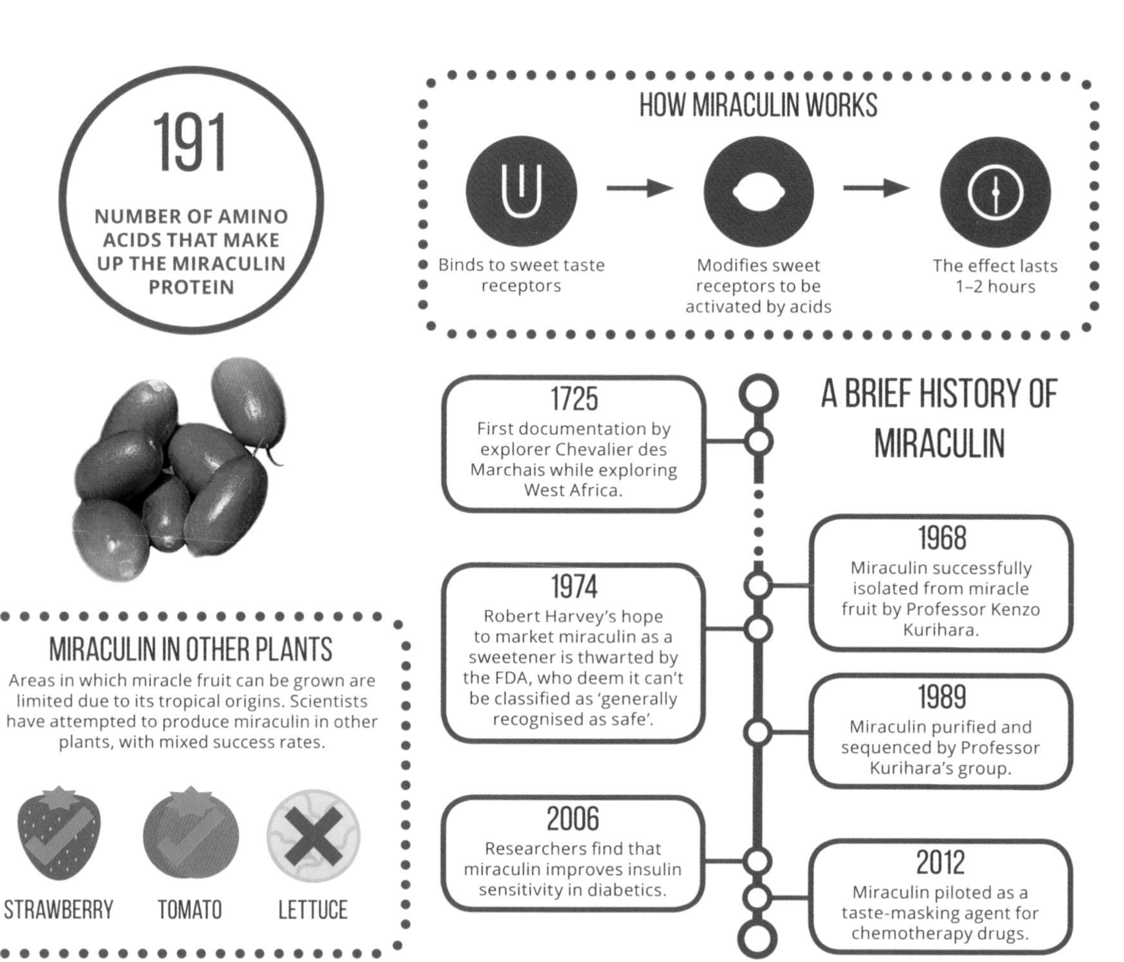
191
NUMBER OF AMINO ACIDS THAT MAKE UP THE MIRACULIN PROTEIN
HOW MIRACULIN WORKS
Binds to sweet taste receptors
Modifies sweet receptors to be activated by acids
The effect lasts 1–2 hours
A BRIEF HISTORY OF MIRACULIN
1725
First documentation by explorer Chevalier des Marchais while exploring West Africa.
1968
Miraculin successfully isolated from miracle fruit by Professor Kenzo Kurihara.
1974
Robert Harvey's hope to market miraculin as a sweetener is thwarted by the FDA, who deem it can't be classified as 'generally recognised as safe'.
1989
Miraculin purified and sequenced by Professor Kurihara's group.
2006
Researchers find that miraculin improves insulin sensitivity in diabetics.
2012
Miraculin piloted as a taste-masking agent for chemotherapy drugs.
MIRACULIN IN OTHER PLANTS
Areas in which miracle fruit can be grown are limited due to its tropical origins. Scientists have attempted to produce miraculin in other plants, with mixed success rates.
STRAWBERRY
TOMATO
LETTUCE

HOW DO MIRACLE BERRIES MAKE SOUR FOODS TASTE SWEET?

Miracle berries (*Synsepalum dulcificum*) grow on a shrub native to West Africa, and have curious taste-modifying abilities. After chewing the fruit, it can make sour food and drink taste sweet for up to an hour after. For example, eating miracle berries before drinking lemon juice removes the sourness completely. This effect is due to the presence of a particular protein which is only found in miracle berries, named miraculin.

As a protein, miraculin has a very large molecular structure – too large to be shown here. To give an idea of its size, it's made up of 191 amino acids, amino acids being the molecular building blocks of proteins. The molecule isn't sweet by itself, and we actually still don't fully understand how it produces its taste-altering effect. Your tongue is covered in receptors capable of detecting different tastes, and what we do know is that miraculin can bind to the taste receptors that detect sweetness very strongly. Then, when you eat or drink something sour, it can react with the miraculin. This reaction has a knock-on effect, causing the shape of the sweet taste receptors to distort. This distortion makes them much more sensitive, and the signals they relay to the brain overpower those of the sour tastes, causing you to experience a sweet taste even with sour foods. This lasts until the miraculin is eventually uncoupled from the receptors.

While this effect might seem like nothing more than a curiosity, food scientists have for many years been looking at ways in which miraculin could potentially be used as a sweetener. There are a number of problems to overcome – firstly, miraculin isn't heat stable, and above 100° Celsius it loses its taste-altering properties, so it isn't possible for it to be used in any foods that require cooking. Another problem is the duration of its effect, though scientists have succeeded in creating a version of the protein whose effect is much shorter lived, which could make its use much easier.

In the USA, the FDA (Food and Drug Administration) has ruled that miraculin is an additive, and cannot be given the status of 'generally recognised as safe', meaning it will likely require several years of testing before it can be included in food products.

WHY DOES ORANGE JUICE TASTE BITTER AFTER BRUSHING YOUR TEETH?

Most people have probably made the mistake of drinking orange juice immediately after brushing their teeth. The effect generated isn't a pleasant one – the resultant taste is bitter, with none of the original sweetness of orange juice. Research suggests that this effect can last up to 30 minutes after the use of toothpaste, and it can also be noticed, in a minor way, on other foodstuffs.

The root of the effect can be found in one of the chemical constituents of toothpaste. Sodium lauryl sulfate is a compound commonly used in personal care products, including toothpaste, shampoo and shower gel. It acts as what chemists call a 'surfactant'. Essentially, the molecule has one end which will dissolve in water and one end which is insoluble in water, but soluble in grease and oils. This helps it dissolve and remove dirt when you wash your hair. It also stimulates foaming; in toothpaste, it lowers the surface tension of saliva, allowing a foam to form. Sodium laureth sulfate is a chemical commonly used instead of sodium lauryl sulfate, but it performs exactly the same role. Some rumours have falsely claimed that these chemicals are cancer causing, but there is scientific evidence that they are completely safe at the concentrations we usually encounter.

The presence of sodium lauryl sulfate (or its alternative) in toothpaste is thought to be the cause of its effect on the taste of orange juice. The commonly accepted explanation for this effect is that sodium lauryl sulfate suppresses the receptors in your mouth that help you detect sweetness. At the same time, it's also thought they can alter membranes formed by compounds called phospholipids, which usually inhibit your bitter taste receptors to an extent. As a consequence, when you drink orange juice immediately after toothpaste, the sweet taste is dulled and the bitter taste is emphasised, producing the somewhat unpleasant effect.

If, for whatever reason, you want to avoid this unfortunate quirk of toothpaste (instead of never brushing your teeth again), there are alternatives to sodium lauryl sulfate toothpastes. These often use glycyrrhizin, a compound isolated from liquorice root, in order to produce their foaming effect.

The surfactant molecules in toothpaste affect our sense of taste.

BITTERNESS

Alters phospholipid membrane structures, which usually inhibit bitterness.

SWEETNESS

Surfactants inhibit sweet taste receptors, reducing sweetness.

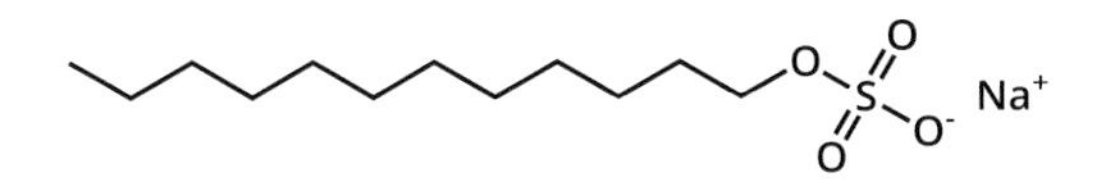

SODIUM LAURYL SULFATE

aka SLS, foaming agent in many toothpastes

SODIUM LAURETH SULFATE

sometimes used in place of SLS

DURATION LASTS FOR UP TO 30 MINUTES

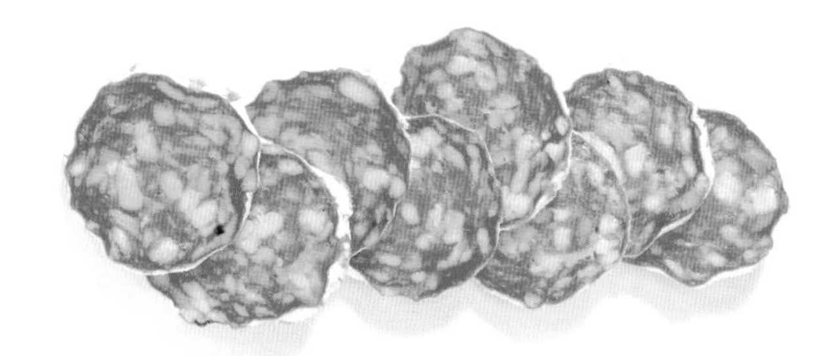

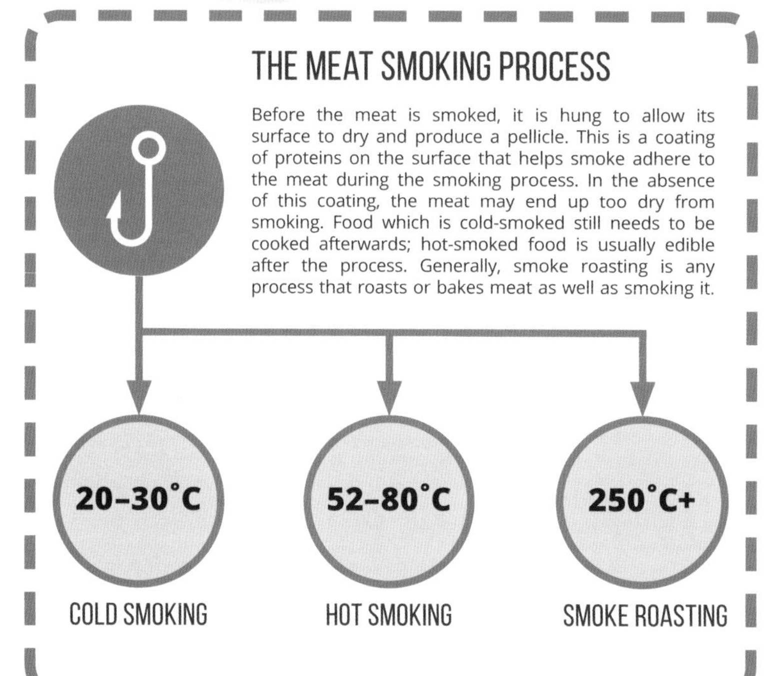

THE MEAT SMOKING PROCESS

Before the meat is smoked, it is hung to allow its surface to dry and produce a pellicle. This is a coating of proteins on the surface that helps smoke adhere to the meat during the smoking process. In the absence of this coating, the meat may end up too dry from smoking. Food which is cold-smoked still needs to be cooked afterwards; hot-smoked food is usually edible after the process. Generally, smoke roasting is any process that roasts or bakes meat as well as smoking it.

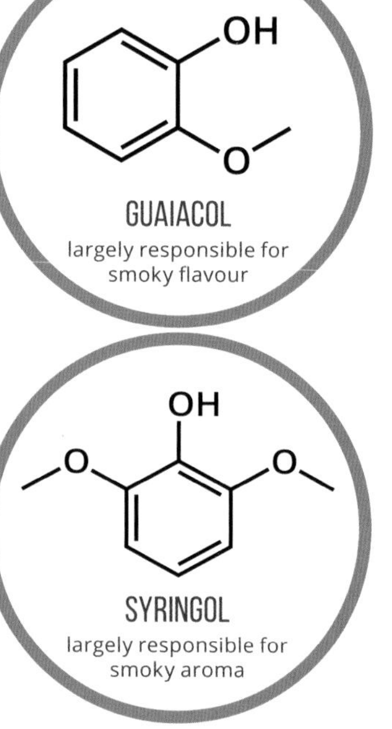

WHY DOES SMOKING MEAT CHANGE ITS FLAVOUR?

Smoking foods, be it ham, bacon, beef or fish, is a cooking method that dates from a time before refrigeration, when the best way to preserve meats that could otherwise start to rot and spoil was to smoke them. Today, we smoke a variety of foods mainly for the enjoyment of the flavour, and there are a number of chemical compounds that contribute towards this.

The process of smoking commonly involves exposing the food to smoke from burning wood. The compounds produced in the smoke are subject to a wide number of factors, such as the type of wood, the temperature and the amount of oxygen. Some compounds have been specifically highlighted as major contributors to the overall flavour and aroma of the meat. The compounds are generated by pyrolysis, which is the thermal decomposition of the organic compounds that make up the wood in the absence of an adequate supply of oxygen.

In general, a class of compounds called phenolic compounds are mostly credited with the specific flavours of smoked foods. One of these, guaiacol, is produced by the breakdown of lignin, a compound which makes up to a third of the dry mass of wood. It's largely responsible for the smoky *flavour* of smoked meats, and is also found in roasted coffee and whisky. It isn't responsible for the smoky *aroma* of smoked meat, however. Another compound, syringol, which is also produced from the pyrolysis of lignin, is the major compound which contributes the smell of smoked food.

So next time you are enjoying a slice of chorizo, remember that a lot of chemical breakdown has taken place to make it so delicious.

WHAT CAUSES THE SOUR TASTE OF GONE-OFF MILK?

Anyone who's accidentally made cereal with gone-off milk can attest to its pretty unpleasant taste and smell. If you're particularly unlucky, the milk will have already reached the stage where small clumps form in solution – not a particularly pleasant addition to your morning cup of tea or coffee.

The reason milk undergoes this change is down to the bacteria that are naturally present. You might think that the whole point of milk being pasteurised is to remove any bacteria, but the main purpose of the process is to remove *harmful* bacteria from the raw milk. After the process, it will still contain the varieties of bacteria that can eventually cause food spoilage. These bacteria gorge themselves on the natural sugar present in milk, lactose, which constitutes about 5 per cent of milk.

As the bacteria feed on lactose, they produce a range of products, one of which is lactic acid. You've probably heard of lactic acid before – it's one of the products of anaerobic respiration, which your body resorts to during exercise when it can't get enough energy from the normal respiration process. In milk, lactic acid contributes the sour, acidic flavour of spoiled milk. It's also responsible for the 'clumping' effect observed.

Casein is the main protein in milk. Normally, the casein molecules repel each other, meaning they float around freely in solution. However, when the milk becomes more acidic, this can cause the molecules to start to clump together, and eventually precipitate out of solution as solid lumps.

Obviously, this is rather inconvenient if it happens to the milk in your fridge. However, the science behind the process is something that's taken advantage of in cheese making. This involves heating milk, then adding acid (e.g. citric acid) in order to make the milk curdle and form clumps. These protein clumps can then be strained and removed from the solution, and eventually used to form cheese.

LACTOSE

Lactose is the main sugar found in milk. It typically constitutes about 5% of milk.

LACTIC ACID

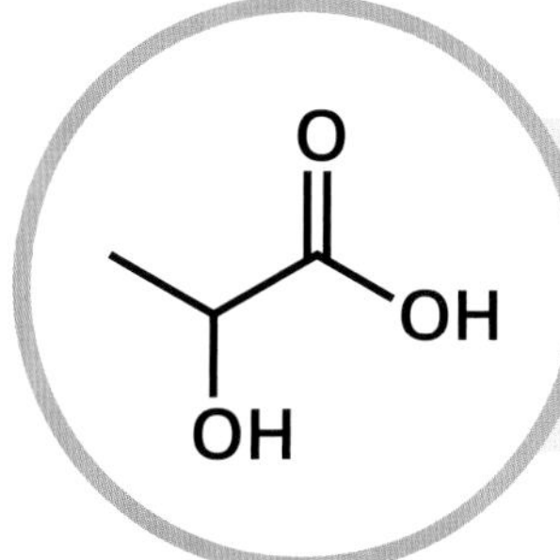

Bacteria in milk break down lactose to form lactic acid, causing a sour taste.

ALPHA CASEIN

One of many casein proteins found in milk

PROTEIN MAKES UP **5%** OF COW MILK

CASEIN ACCOUNTS FOR **80%** OF MILK PROTEIN CONTENT

82%

Percentage of aldehyde compounds in the essential oil of coriander leaves

2-decenal is also a compound found in the secretions of stink bugs

Aldehydes are amongst the by-products of the soap-making process

ALDEHYDES IN CORIANDER

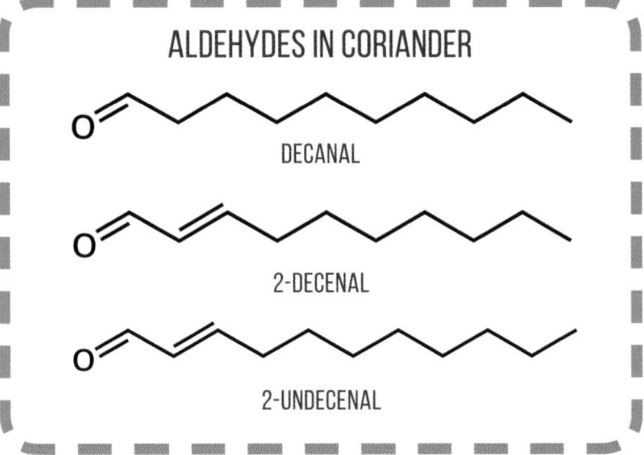

A GENETIC BASIS?

A dislike of the taste of coriander has been linked with genes for taste and smell – thought it is unlikely that the dislike is solely as the result of genetics.

WHY DOES CORIANDER TASTE SOAPY TO SOME PEOPLE?

There are quite a few people for whom coriander has a rather unpleasant soapy, or even metallic, taste. The cause of this has its roots in the chemical composition of coriander leaves, but there can also be other factors at play that determine whether or not you're a fan of coriander.

The chemical composition of the essential oil of coriander leaves has shown it to be composed of around 40 different organic compounds, with 82 per cent of these being aldehydes, and 17 per cent alcohols. The aldehydes are mainly those with 9–10 carbon atoms, and it is these that are largely responsible for the aroma of coriander leaves – as well as its perceived soapy taste for some people.

The aldehydes present in coriander, as well as those similar to them, are also commonly found in both soaps and lotions. Interestingly, some are also amongst the compounds excreted by shield bugs (also known as stink bugs) when they are disturbed. Given that, perhaps it's not completely surprising that, for some people, the smell and taste of coriander is a little on the repulsive side.

However, it's certainly not just down to the chemical composition of coriander leaves that makes some find it has a soapy taste. It's been suggested that there's also a genetic basis to this, which explains why not everyone has the same aversion. Scientists have highlighted a specific gene that codes for a receptor that is highly sensitive to the flavour of aldehydes. Several other genes have also been linked, however, so it seems likely that more than one could be responsible.

As well as this, it's also possible for people to grow to like the taste of coriander, with it being suggested that repeated exposure to the taste leads to the brain forging new, positive associations. The strength of the aldehydes' effect on the taste of coriander can also be mitigated by crushing the leaves before consumption, with studies having shown that this speeds up the rate at which the aldehydes in the leaves are broken down by enzymes.

WHAT DO DILL AND SPEARMINT HAVE IN COMMON?

Dill and spearmint – two herbs that don't have a whole lot in common besides being green herbs. Dill has a mild, sweet taste and fragrance, while spearmint has a minty aroma and is often used as the flavouring in mint toothpastes. Despite these rather marked differences, the same compound, carvone, is responsible for the flavour and smell of both. Carvone can exist as two individual optical isomers, and these two isomers have discrete characteristics.

First of all, if you don't have a chemistry background, you might be wondering what we mean by the term 'isomer'. An isomer of a molecule is a compound with the same molecular formula – that is, the same number of each type of atom – but in a different arrangement. There are different types of isomers, and optical isomerism is a specific type where the two isomers are mirror images of each other. The two mirror images, or 'enantiomers', cannot be superimposed on one another. Your hands are a perfect model of non-superimposable mirror images: it's impossible for you to place your left hand on your right hand so it looks exactly the same, and similarly, it's impossible to place one optical isomer on top of another so all of the atoms are in the exact same place.

So, the two optical isomers of carvone have the same molecular formula, but the isomer which smells and tastes like dill is the mirror image of that which smells and tastes like spearmint. Why does this seemingly minor difference in the arrangement of atoms result in very different smells and tastes? While we still don't fully understand the intricacies of our sense of taste and smell, they can be considered along the lines of a lock and key model. The taste or smell receptors are the lock, and specific molecule 'keys' are required to activate each. Hence, one optical isomer of carvone activates certain smell and taste receptors, while the other optical isomer, being a mirror image, activates others.

Optical isomerism is also important in medicine, the famous case being thalidomide – a morning sickness drug prescribed in the 1950s and 60s. One optical isomer was beneficial and the other caused tragic birth defects.

S-CARVONE

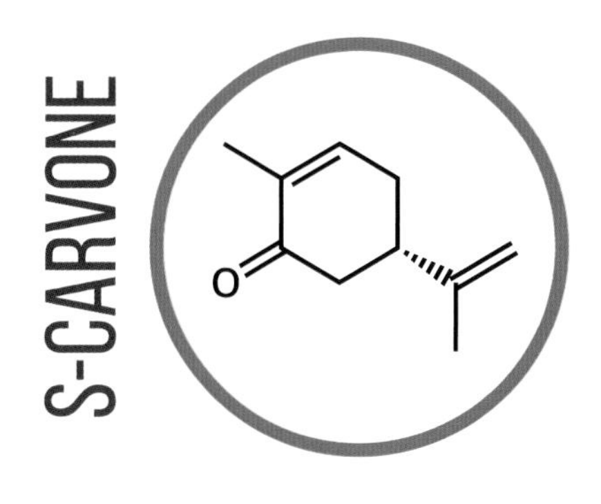

R-CARVONE

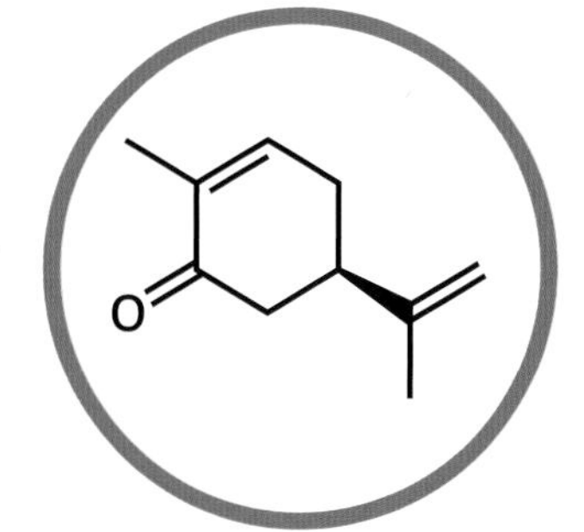

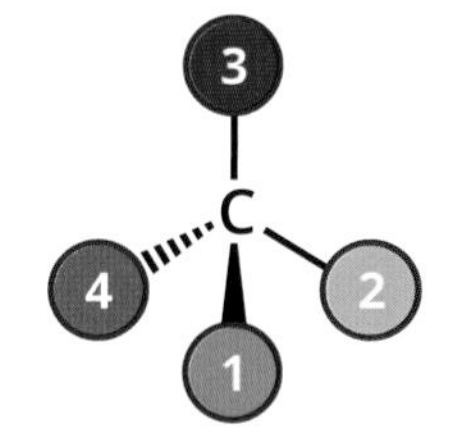

(S) ENANTIOMER

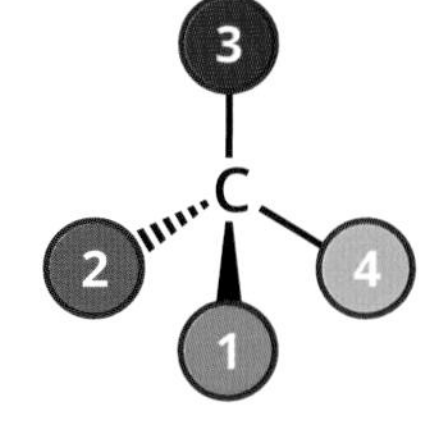

(R) ENANTIOMER

OPTICAL ISOMERISM

Many organic compounds have optical isomers. Their physical properties, such as melting and boiling point, are identical. They often differ in benign ways, for example in the smell or flavour of the compound. However, some pharmaceuticals with optical isomers can show different, unwanted effects with one isomer. An example is that of thalidomide.

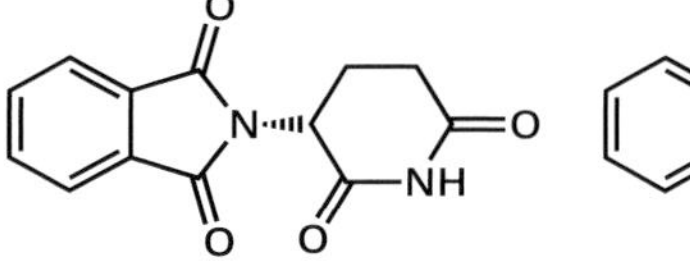

S-THALIDOMIDE
(teratogenic)

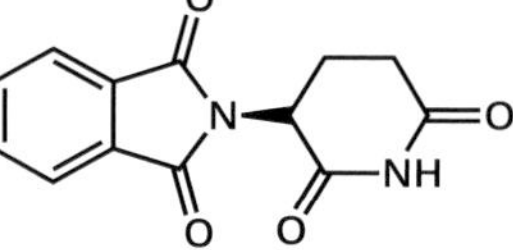

R-THALIDOMIDE
(sedative)

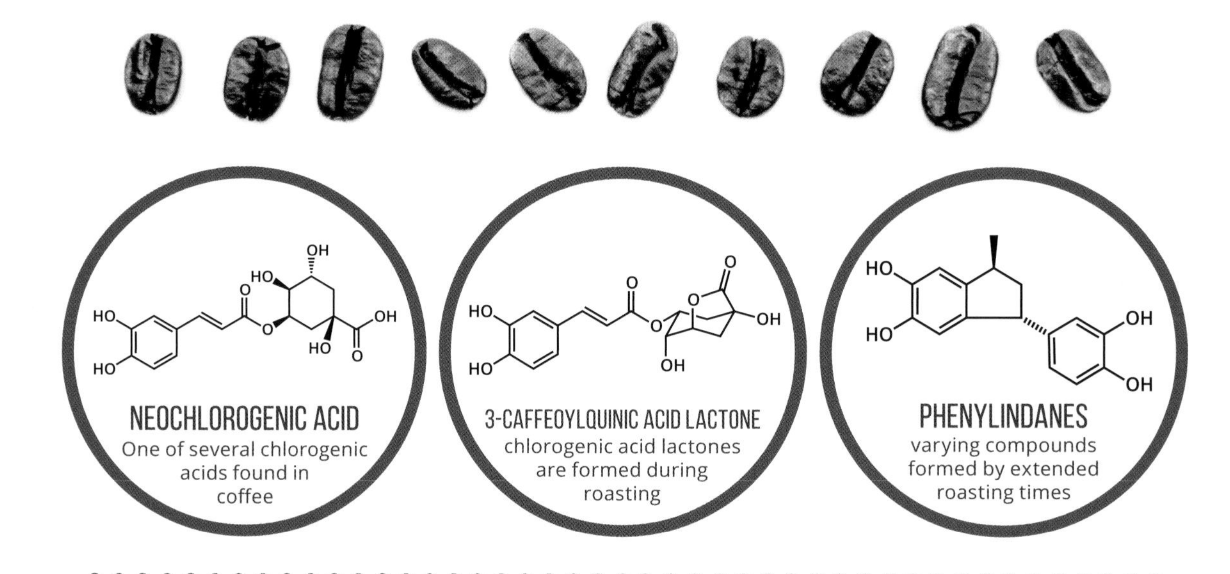

Light roast

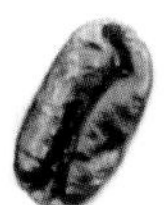

Medium roast

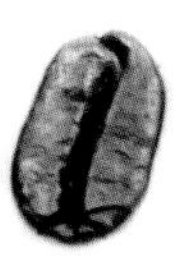

Dark roast

Main source of bitterness: chlorogenic acid lactones

Main source of bitterness: phenylindanes

WHAT CAUSES THE BITTERNESS OF COFFEE?

Mention coffee, and caffeine is the chemical compound name that immediately springs to mind. Caffeine is responsible for the stimulant effect of coffee. It binds to receptors in the brain to which the brain chemical adenosine would usually bind. Adenosine's effect on binding is to make you feel tired – caffeine stalls this process by blocking adenosine's access to the receptors. The effects of caffeine are well documented, but it contributes relatively little to the bitter taste of coffee. This is where the other compounds in coffee beans come in.

Coffee, as it turns out, is a cornucopia of chemical compounds that influence its taste. One group on which a lot of research has been carried out are the chlorogenic acids. These compounds account for up to 8 per cent of the composition of unroasted coffee beans. When coffee beans are roasted, the chlorogenic acids react to form a variety of different products, which can all affect the taste of the coffee.

In medium to light coffee brews, the main source of bitterness is from the formation, from chlorogenic acids, of compounds called chlorogenic acid lactones. With longer roasting times, these too break down. In dark roasted coffees, the breakdown products of the chlorogenic acid lactones have an increasing effect on the bitterness of the flavour. These products are called phenylindanes, and their bitterness is harsher than that of the chlorogenic acid lactones – explaining, for example, the bitterness of espresso coffee. Research has found that the brewing method utilised could also be a factor in perceived bitterness, with espresso-style brewing methods at high temperature and pressure yielding a higher concentration of bitter compounds than other methods. The role of chlorogenic acids in the flavour of coffee has been debated by some, and research is still ongoing.

WHAT CAUSES THE BITTERNESS AND TASTE OF BEER?

Chances are, when you crack open a beer, you probably won't be thinking much about chemistry – but it's the particular chemicals in beer, produced in the brewing process, that give beer both its bitterness and flavour.

Compounds originating from hops account for the bitter flavour. Hops contain organic compounds called alpha and beta acids. There are five main alpha acids: humulone, cohumulone, adhumulone, posthumulone and prehumulone. During the brewing process, they are degraded to form iso-alpha acids which are more soluble, and they cause much of the bitter taste. Hops with varying compositions may be selected in order to vary the type and level of the bitterness in the beer.

There are three main types of beta acid compounds, lupulone, colupulone and adlupulone. They impart a harsher bitterness than the alpha acids, but as they are insoluble their contribution is much lower. They do not isomerise in the same way as the alpha acids during fermentation, but instead slowly oxidise to produce their bitter flavour. Because they take much longer to do this, their effects become more potent the longer the beer is fermented and aged.

Essential oils from the hops are responsible for the bulk of the aroma and flavour. Some of these oils are very volatile, evaporating easily; for this reason, they are usually obtained by adding hops later in the brewing stage, or by using 'dry hopping', a technique which involves soaking hops in the finished beer for several days or weeks.

Over 250 essential oils have been identified in hops. Of these, myrcene, humulene and caryophyllene are the main oils found in the highest concentrations; humulene, in particular, is responsible for the characteristic hoppy aroma of beer. American hop varieties tend to have higher amounts of myrcene, which adds a citrus or piney aroma, while caryophyllene contributes a spicy flavour.

A final class of compounds, esters, can also play an important part in the flavour. They are present in varying degrees, dependent on the type of beer – lagers will contain minimal concentrations, while in ales they are much higher. They form through the reaction of the organic acids in hops with the alcohol in beer, along with a molecule called acetyl coenzyme (also found in hops). As volatile flavour compounds, they are responsible for a fruit-like taste.

Different esters lend different smells. Ethyl acetate is one of the most common – it actually has an aroma not dissimilar to that of nail polish at high concentrations, but at the concentrations in beer it imparts a fruity aroma. Isoamyl acetate gives a banana-like scent, ethyl butanoate gives an odour described as tropical or reminiscent of pineapple, and ethyl hexanoate an apple and anise note.

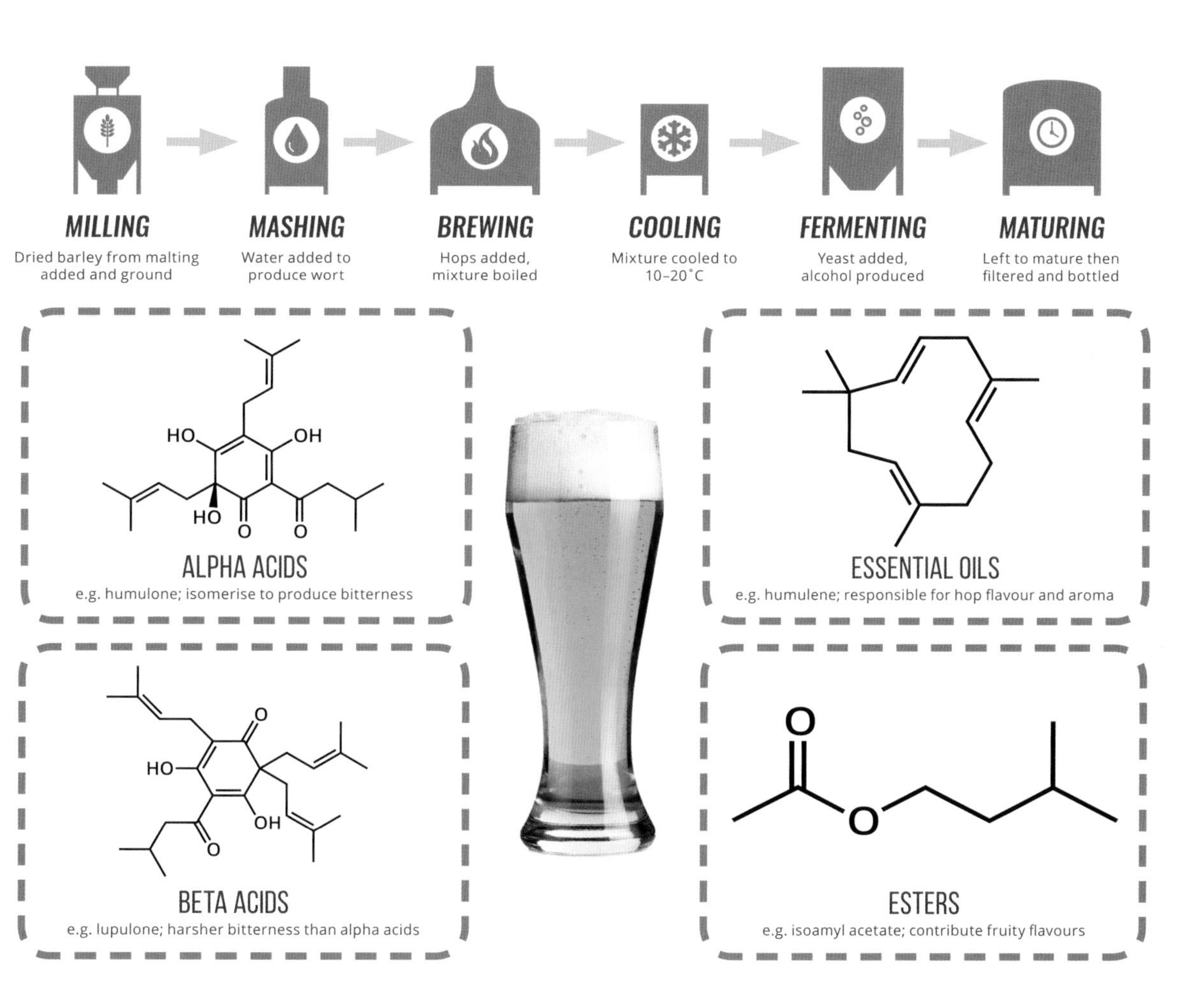
MILLING
Dried barley from malting added and ground
MASHING
Water added to produce wort
BREWING
Hops added, mixture boiled
COOLING
Mixture cooled to 10–20˚C
FERMENTING
Yeast added, alcohol produced
MATURING
Left to mature then filtered and bottled
HO
OH
HO
O
O
ALPHA ACIDS
e.g. humulone; isomerise to produce bitterness
ESSENTIAL OILS
e.g. humulene; responsible for hop flavour and aroma
O
HO
OH
O
BETA ACIDS
e.g. lupulone; harsher bitterness than alpha acids
O
O
ESTERS
e.g. isoamyl acetate; contribute fruity flavours

AROMA

WHY DOES GARLIC MAKE YOUR BREATH SMELL?

Garlic is frequently used in cooking, but its use comes with the inevitable and unwanted accompaniment of 'garlic breath'. Much as with onions, the chemicals that cause this effect aren't actually present in unchopped garlic, but are formed once it is chopped.

When garlic cloves are mechanically damaged, the enzyme alliinase, usually kept segregated within the cells, is released. This enzyme breaks down the chemical alliin, found in the cloves, to form allicin. This is actually a part of garlic's natural self-defence mechanism to protect it from insects and fungi. Allicin is the major compound that contributes to chopped garlic's aroma. It's quite unstable, and is subsequently itself broken down into a range of sulfur-containing organic compounds, several of which contribute to the 'garlic breath' effect.

Research has identified four major compounds that contribute: diallyl disulfide, allyl methyl sulfide, allyl mercaptan and allyl methyl disulfide. While some of these are broken down within the body quickly, others linger. Allyl methyl sulfide is the compound that takes longest for the body to break down. It is absorbed in the gastrointestinal tract and passes into the bloodstream, then passes on to other organs in the body for excretion, specifically the skin, kidneys and lungs. It is excreted through the skin via sweating, in the urine – and through your breath. This effect can last up to 24 hours, until all of the compound is excreted from the body, causing a faint, lingering, garlicky aroma.

So, what can you do to mitigate this effect? Some research has been carried out in the area, and a number of foods have been discovered to mildly reduce garlic breath. These include parsley, milk, apple, spinach and mint. Try eating them if you need to freshen up.

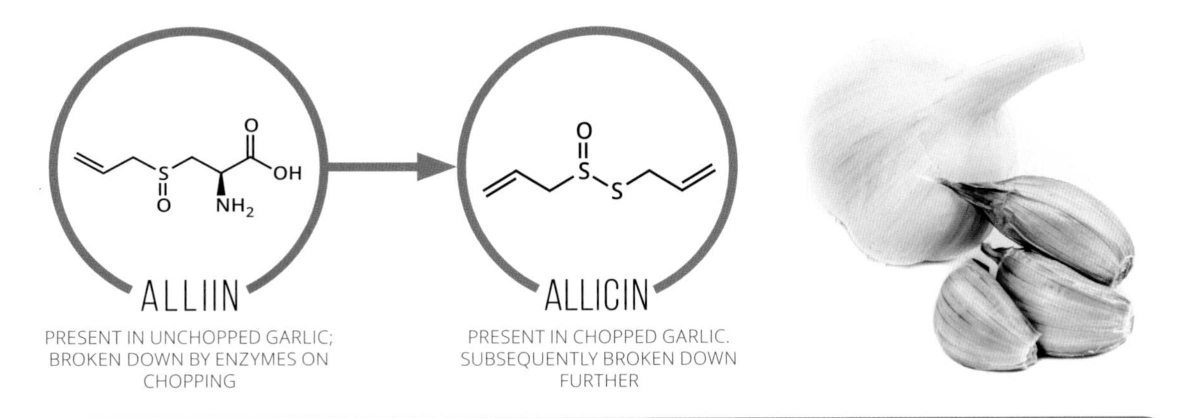

AFTER FURTHER ENZYMATIC BREAKDOWN, THE FOLLOWING COMPOUNDS ARE PRODUCED

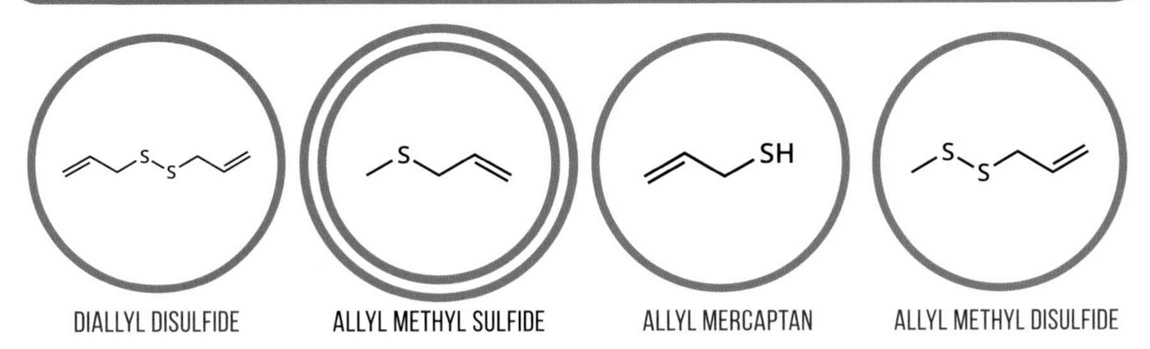

ALLYL METHYL SULFIDE IS BROKEN DOWN IN THE BODY MORE SLOWLY, AND IS THE MAIN CAUSE OF GARLIC BREATH

It is excreted from the body via the lungs, skin and urine. The effect can last for up to 24 hours!

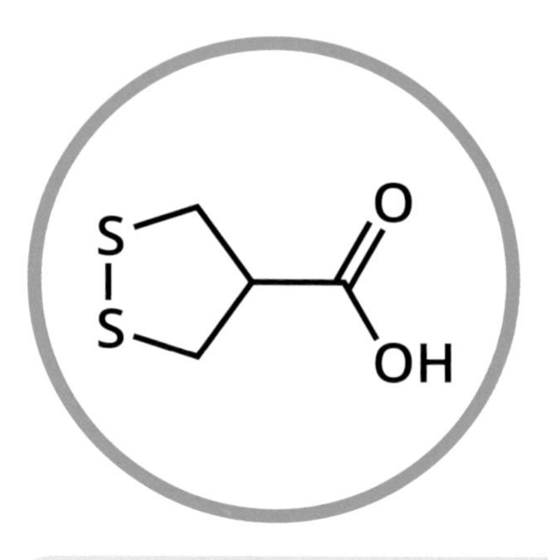

ASPARAGUSIC ACID
(FOUND ONLY IN ASPARAGUS)

ASPARAGUSIC ACID BREAKDOWN PRODUCTS

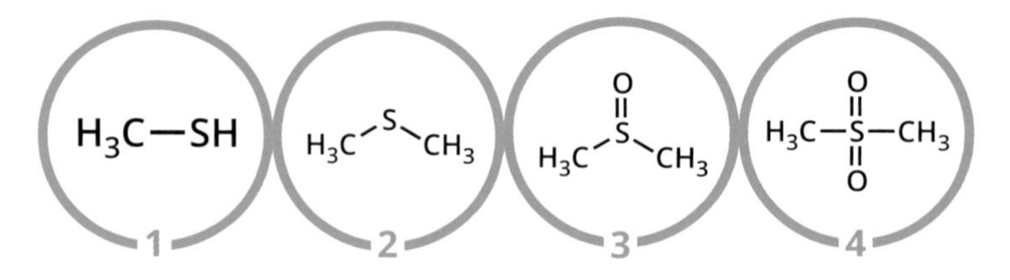

1. METHANETHIOL
2. DIMETHYL SULFIDE
3. DIMETHYL SULFOXIDE
4. DIMETHYL SULFONE

ASPARAGUSIC ACID
↓
SULFUR-CONTAINING COMPOUNDS
↓
UNPLEASANT AROMA

WHY DOES ASPARAGUS MAKE YOUR WEE SMELL?

If you've ever eaten asparagus, you may well have noticed that, a short time after ingestion, it imparts a strange, somewhat unpleasant scent to your urine. But then again, you may have noticed nothing of the sort – and there's a scientific explanation for that too.

The chemicals that cause the effect are all suspected to stem from just one chemical compound: asparagusic acid (which is found naturally only in asparagus – hence its name). It has been pinpointed as the probable source of several organic compounds that have been shown to affect the odour of urine.

When we eat asparagus, the asparagusic acid molecules contained therein are broken down by digestion into a number of sulfur-containing organic compounds. In studies, a technique known as gas chromatography-mass spectrometry was used to analyse the 'headspace' of urine produced after consumption of asparagus. The headspace is the gas space immediately above the liquid surface, which is occupied by light, volatile compounds in the liquid, and analysis of this is useful in identifying odour-causing compounds. The analysis of the post-asparagus urine showed the presence of several compounds that were not present, or present in negligible amounts, in normal urine. The primary compounds present, in quantities a thousand times greater than in normal urine, were methanethiol and dimethyl sulfide. The compounds dimethyl sulfoxide and dimethyl sulfone were also present, and it was suggested that they modify the aroma to give it a 'sweet' edge.

The human nose is very sensitive to thiol compounds – a concentration as low as a few parts per billion is enough for us to be able to detect them. To give you an idea of how bad thiol compounds can smell, they're also found in skunk spray. So, the increase in concentration of these compounds in urine after eating asparagus goes a long way towards explaining why the effect is so potent. The odour is detectable remarkably quickly after eating asparagus, within 15 to 30 minutes.

Interestingly, the ability to smell the aroma of asparagus-influenced urine is not universal. Research has shown that a proportion of people are unable to detect the change in smell, with one study finding 2 out of 31 people were unable to detect a difference in odour after eating asparagus. It was initially thought that everyone produced the odour, but only some could smell it; however, it has since been suggested, after a range of research, that not all people exhibit the effect after eating asparagus, with another study placing the figure of people who do produce 'asparagus urine' at 43 per cent.

WHY DOES DURIAN FRUIT SMELL SO BAD?

The durian originally hails from South East Asia, and is grown in countries including Thailand, Malaysia, Vietnam and Indonesia. This fruit is distinguished by its large, thorny appearance – and its absolutely horrendous odour. The taste of the flesh is pleasant and creamy, but the smell has been described as a delicious-sounding blend of onion, cheese and rotting meat. It's so bad it has been banned on Singapore's rail network and in a large number of airports in the region.

A potent blend of volatile organic chemicals causes the durian's offensive odour. These can be ordered by their 'flavour dilution' (FD) factor – a measure of how much they have to be diluted before their smell can no longer be detected. Essentially, the higher the FD factor, the smellier the compound. A 2012 study found a total of 50 different organic compounds that contributed to the smell of the durian, and gave the approximate FD factors for each.

Amongst the highest were:

- ethyl-(2S)-2-methylbutanoate, FD factor 16384, fruity odour.
- ethyl-cinnamate, FD factor 4096, honey odour.
- 1-(ethylsulfanyl)ethanethiol, FD factor 1024, roasted onion odour.

Some other notable compounds include the 'skunky' 3-methylbut-2-ene-1-thiol; hydrogen sulfide, with its characteristic smell of rotten eggs; propane-1-thiol, whose odour is actually described as 'rotten/durian'; and a number of other sulfur-containing organic compounds, whose odour is perhaps unsurprisingly described as 'sulfury'.

However, if you can get past the terrible smell, a durian could pose you other problems – particularly if you wash it down with any alcohol-containing drinks. It's been suggested that the sulfur compounds in the durian can interfere with aldehyde dehydrogenase (ALDH), an enzyme responsible for breaking down acetaldehyde in the body. Acetaldehyde is a compound produced by the metabolism of alcohol in the body, and one we'll talk about in more detail later in the book. Durian can inhibit its breakdown by as much as 70 per cent, causing it to build up in the body before it, too, can be broken down.

A final danger that a durian poses is merely in its size and shape. It's a hefty fruit, and with a tough shell of spines has been known to fall off trees and land on farmers, causing potentially serious injuries.

50 Compounds that contribute to a durian's smell

Banned on mass transit in Singapore

Commonly called 'king of fruits' in South East Asia

Described as a 'love it or hate it' flavoured fruit

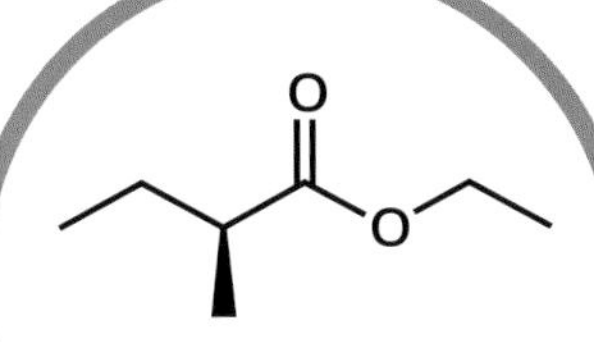

ETHYL-(2S)-2-METHYL BUTANOATE
'fruity' odour

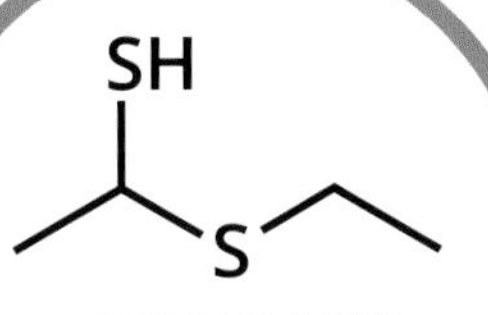

1-(ETHYLSULFANYL) ETHANETHIOL
'roasted onion' odour

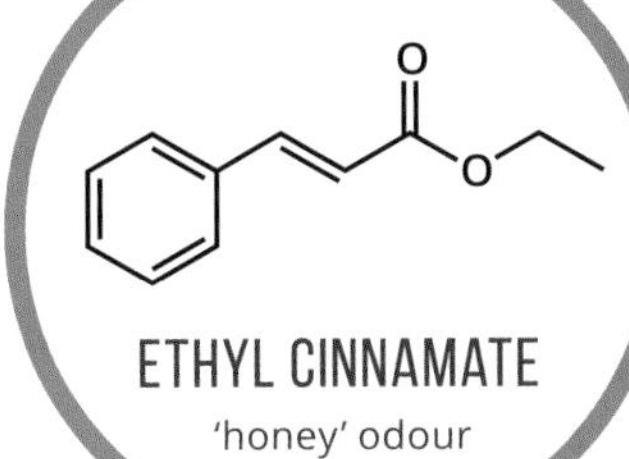

ETHYL CINNAMATE
'honey' odour

SH

3-METHYLBUT-2-ENE-1-THIOL
'skunky' odour

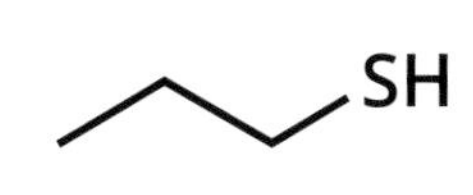

PROPANE-1-THIOL
'rotten' or 'durian' odour

PYRAZINES, FURANS & PYRIDINES

Thought to be responsible for the characteristic meaty aromas in bacon

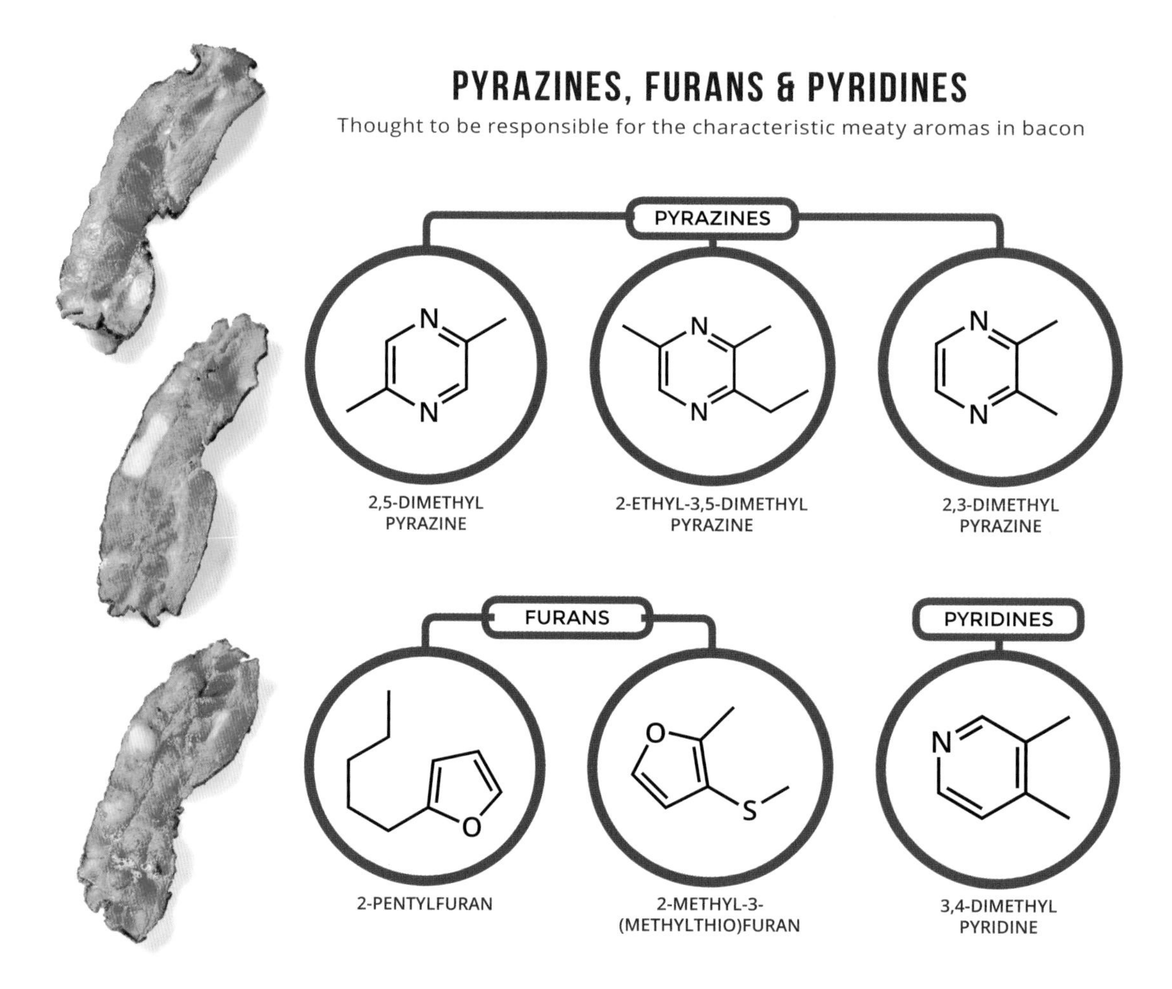

WHY DOES BACON SMELL SO GOOD?

When it comes to breakfasts, there are few aromas better than that of bacon sizzling and crisping in a pan. This delicious aroma is the consequence of specific volatile chemical compounds given off as it is frying.

Despite being such a worshipped food item, research on the compounds behind the aroma of frying bacon is surprisingly sparse. In fact, at the time of writing there seems to be only one study, from 2004, that focuses specifically on the aroma-causing compounds in bacon. In this research, scientists tried to discover the compounds that give frying bacon its aroma by comparing it with the aromas released when cooking pork loin. They did this by frying the meat, mincing it, then passing nitrogen gas over it to collect any volatile organic compounds that were being released.

The volatile compounds collected originate in part from the Maillard reaction, in which sugars in the foodstuff are broken down by reaction with amino acids as heating occurs. In bacon, other volatile compounds are produced due to the thermal breakdown of fat molecules. As well as this, in the case of smoked bacon, nitrite used in the curing process can also react with the fatty acids and fats present in bacon on heating – this leads to a higher percentage of nitrogen-containing compounds than in standard pork meat.

So, what compounds give bacon its aroma? The researchers compiled an exhaustive list of the volatile compounds present; they found that hydrocarbons, alcohols, ketones and aldehydes were present in large quantities in both the bacon and pork aromas, though not all of them are necessarily contributors to the perceived scent. However, they found some compounds present exclusively in bacon, and suggested that these play a major role in its aroma.

These were all nitrogen-containing compounds; they included 2,5-dimethylpyrazine, 2,3-dimethylpyrazine, 2-ethyl-5-methylpyrazine and 2-ethyl-3,5-dimethylpyrazine. The researchers found that, individually, none of these compounds had the precise smell of bacon – however, they suspect that combined, and in combination with some other volatile compounds, it is most likely that they are responsible. As well as these compounds, others that had previously been identified from other meats as having a 'meaty' scent were isolated. These included 2-pentylfuran, an oxygen-containing organic compound, and 3,4-dimethylpyridine, another nitrogen-containing compound.

This research, then, gives us a general idea of some of the contributing compounds. It's likely that it's not the full picture, and more studies would probably be needed to identify the precise combination of compounds responsible for bacon's heady aroma. It also doesn't factor in the human response. This variance from person to person is something that's harder to quantify chemically. Hard as it may be to believe, to some people bacon might not smell that great!

WHAT CAUSES THE SMELL OF FISH?

While fish can taste delicious, their fishy odour, which intensifies as the fish becomes less fresh, is a bit of a turn off for some. It also has a tendency to cling to your fingers if you've been handling fish, as anyone who's ever worked on a fish counter or prepared fish will attest. So what causes this formidable scent?

The chemical compound that causes the smell actually stems from the natural habitat of salt water fish. On average, sea water contains about 35 grams of salt per litre. Osmosis is the process of water molecules moving from a high concentration to a lower concentration, so in order to maintain their cell water levels, fish cells will contain substances called osmolytes, which are soluble in the cells and help maintain cell volume. In fish, the main osmolyte is a compound called trimethylamine oxide.

Trimethylamine oxide itself is odourless. However, when a fish is caught, killed and removed from the sea, a combination of enzymes and bacteria contrive to break this chemical down. Amongst the breakdown products is trimethylamine, and it's this compound that contributes the smell we associate with fish. It's also a good indicator of how fresh the fish is – the more trimethylamine and the fishier the fish smells, the longer it has been out of the ocean. A fish from the ocean shouldn't actually smell 'fishy' at all immediately after it's caught! Freshwater fish don't smell anywhere near as fishy, since they contain much lower levels of trimethylamine oxide.

What can be done to reduce the smell of fish, particularly if you want to remove the odour from your hands after preparing a meal? Some basic chemistry is able to help you out here. Amines like trimethylamine are alkaline substances – which means that they can be neutralised by acids. Lemon juice is the most commonly suggested remedy, but any other acidic foodstuff could theoretically be used.

If you can't stand the smell of fish, spare a thought for sufferers of fish malodour syndrome, or trimethylaminuria to give it its medical name. This is a rare disorder which causes production of a particular enzyme in the body to be defective. People with a defective version of this enzyme are unable to oxidise trimethylamine back to trimethylamine oxide in the body, and so it accumulates and is excreted from the body via sweat, urine and breath – literally making the unfortunate sufferer smell of fish.

BACTERIA & FISH ENZYMES

Break down TMAO into a number of compounds.

Trimethylamine content is a common way of assessing fish freshness.

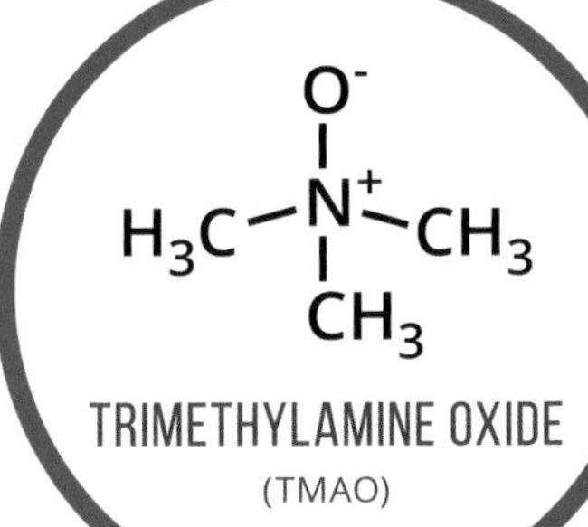

TRIMETHYLAMINE OXIDE

(TMAO)

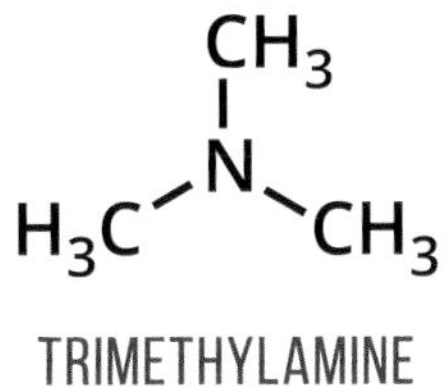

TRIMETHYLAMINE

main cause of the odour of fish

THE 'MUDDY' TASTE OF FRESHWATER FISH

Freshwater fish contain much lower levels of trimethylamine oxide, and as such do not exhibit such a fishy smell. They can, however, have a muddier flavour, caused by presence of the compounds geosmin and 2-methylisoborneol.

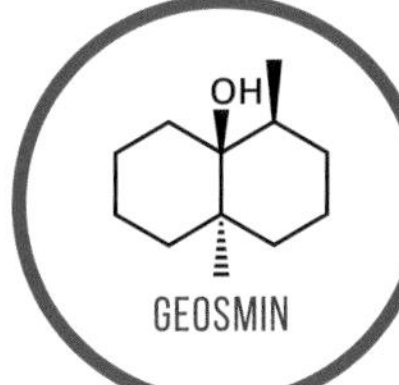

GEOSMIN

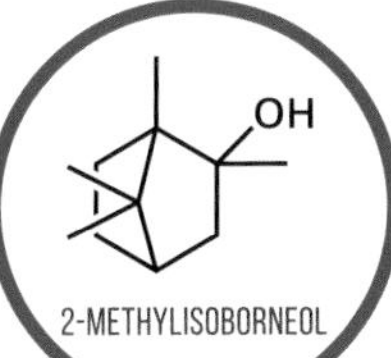

2-METHYLISOBORNEOL

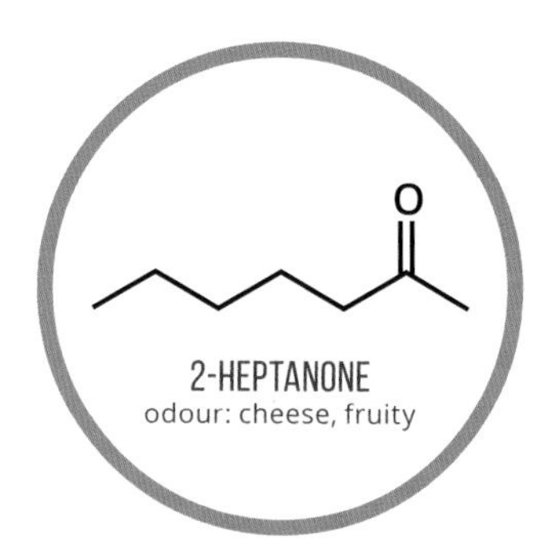

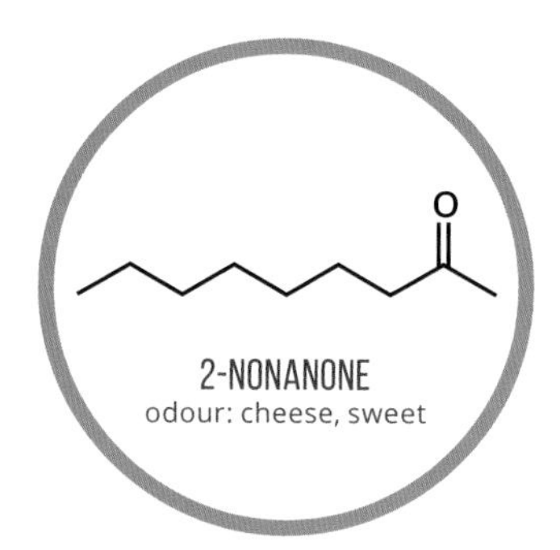

THE VARYING ODOURS OF DIFFERENT BLUE CHEESES

Blue cheeses smell differently, due to varying concentrations of the odour-causing chemicals in the cheese. Some of the main odour-causing chemicals in different cheeses are shown below; bold text indicates major components.

STILTON	GORGONZOLA	ROQUEFORT
2-HEPTANONE	**2-NONANONE**	**2-HEPTANONE**
2-BUTANONE	2-HEPTANONE	**2-NONANONE**
2-PENTANONE	2-UNDECANONE	2-PENTANONE

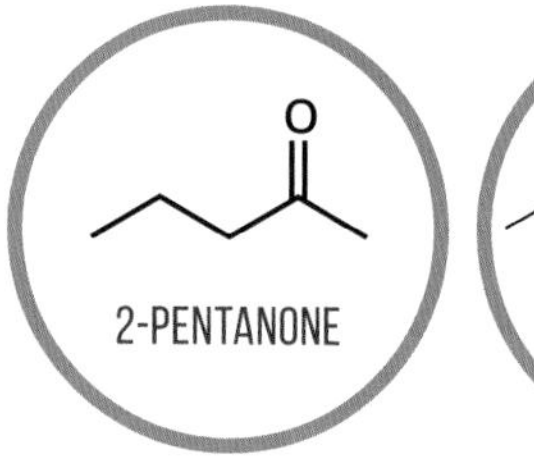

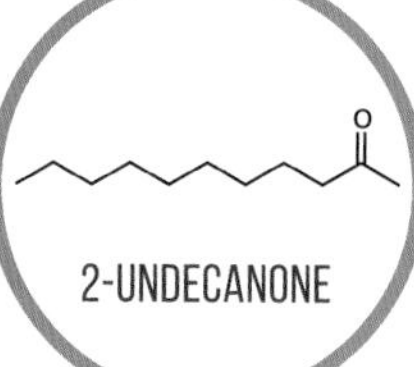

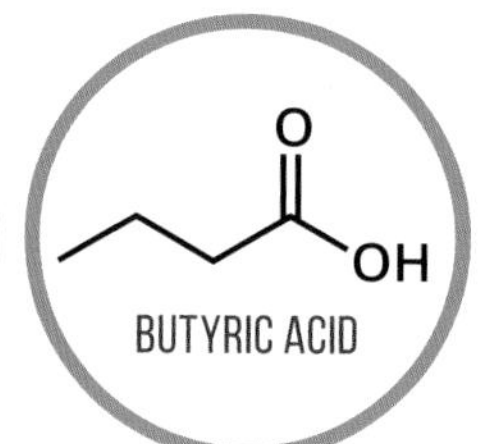

OTHER CHEESE ODORANTS

A number of other chemicals can also contribute to cheese odour. Some are also found elsewhere; for example, butyric acid is a constituent in the smell of human vomit.

WHAT CAUSES BLUE CHEESE TO SMELL SO STRONG?

Of all the cheeses, blue cheese is the type with the most distinctive odour. There's certainly no mistaking the aroma of a block of Stilton or Roquefort, but what is it that makes the smell of these cheeses so much stronger than others, such as Cheddar?

Blue cheese owes its differing appearance and aroma to the type of mould purposefully added during the cheese-making process. The particular moulds added are penicilliums, so called because they produce penicillin. One of the most commonly used is penicillium roqueforti, named after the French town of Roquefort, and unsurprisingly the mould added to produce Roquefort cheese, although it's also used to make Danish blue and Stilton. The mould is introduced to the cheese, then the cheese is pierced or 'needled' to allow air to enter. By feeding the mould, this encourages the formation of the characteristic green and blue veins that run through blue cheese.

As it grows, the mould produces enzymes which convert fatty acids in the cheese into a class of compounds called n-methyl ketones. There are a wide number of different members of this family of compounds that can potentially be produced, but the most important as far as taste and smell are concerned are 2-heptanone and 2-nonanone. Both of these compounds have odours independently described simply as 'blue cheese'.

Obviously, in different blue cheeses, these compounds are present in varying amounts. They're the most abundant in Roquefort, and 2-heptanone is the most abundant in Stilton, with 2-nonanone the most abundant in Gorgonzola. 2-pentanone is also present in relatively large amounts in Stilton, and has an aroma described as 'malty and fruity'. So that's why all blue cheese doesn't smell exactly the same.

One final mention on the smell of cheese, though, has to go to Parmesan. One of the key aroma chemicals in Parmesan cheese is butyric acid. This compound leads a rather unfortunate double life, in that it's also one of the key contributing compounds to the smell of human vomit. Interestingly (or perhaps disgustingly), blindfolded test subjects found the smell of a mixture of butyric acid and another compound, isovaleric acid, appealing when told it was the smell of Parmesan – but revolting when told what they were smelling was the smell of vomit.

WHY DO BEANS GIVE YOU FLATULENCE?

The flatulence-inducing properties of beans need no introduction; it's well known that eating even a small portion can have unwelcome consequences. The reason for this is down to their chemical composition, and what happens to these chemicals in our gut once we've eaten them.

Beans contain a particular type of sugars called oligosaccharides. These are a type of polysaccharide – molecules formed of long chains of saccharides, or sugars. (Examples of monosaccharides are fructose, glucose and dextrose.)

Particularly abundant oligosaccharides in beans are raffinose and stachyose. Because they are such large molecules, they are particularly resistant to digestion; the enzymes in our digestive system responsible for breaking down foods aren't able to make them small enough so that they can be absorbed through the walls of the small intestine. So, when you eat a portion of beans, they are likely to make it to your large intestine without having been split up into smaller portions.

When they do reach the large intestine, they encounter the huge number of natural bacteria that inhabit it – what's sometimes referred to as the 'gut flora'. These bacteria are more than happy to do the job that our digestive system can't, and break down any oligosaccharides that come their way. As they do so, their activity produces a range of gases, which include carbon dioxide and hydrogen. These can eventually lead to the production of gases such as hydrogen sulfide (as well as methanethiol and dimethyl sulfide), a particularly unpleasant-smelling chemical commonly found in human flatulence. Beans aren't the only vegetables which can cause this effect; onion, garlic, cauliflower, cabbage and Brussels sprouts are amongst the many others that contain oligosaccharides or polysaccharides that our digestive system has trouble with.

So, how can the effect of beans be reduced? Well, one suggestion is soaking the beans prior to cooking. This might not work so well with things like baked beans which already come in tomato sauce, but for other types it can help eliminate some of the oligosaccharides. However, it only reduces them by around 25 per cent, so it's not going to prevent the problem completely. Alternatively, a dietary supplement called Beano can be taken before eating them. It contains an enzyme that helps break down oligosaccharides and polysaccharides before they reach the large intestine.

As it turns out, oligosaccharides may not even be entirely to blame. It's been suggested that the proteins and polysaccharides that act as cell wall cements in the plant cells can also pose a similar problem to our digestive systems, and lead to the production of gas in a similar manner.

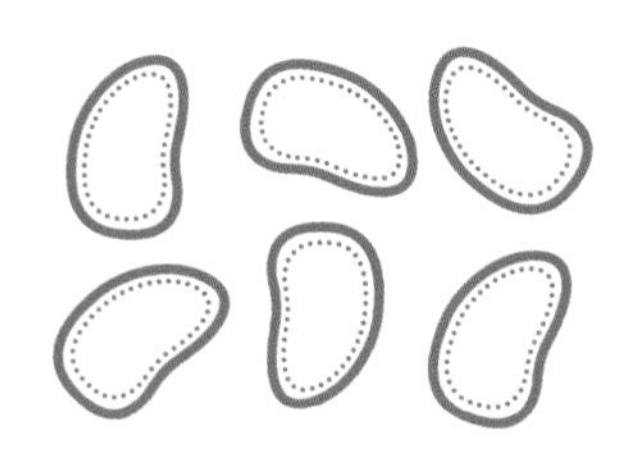

BACTERIA

Break down oligosaccharides that the body is unable to, producing volatile sulfur by-products.

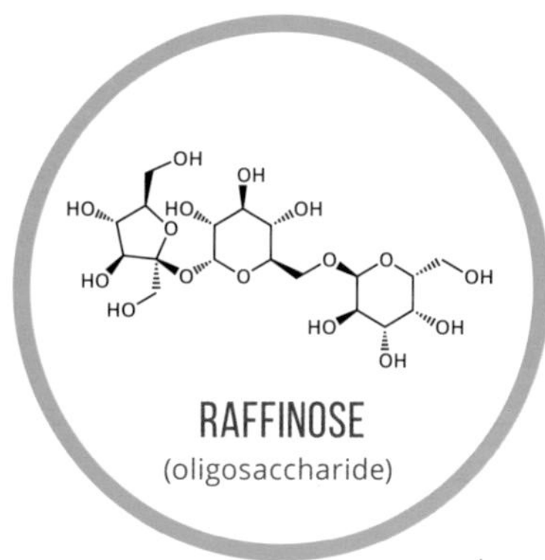

RAFFINOSE

(oligosaccharide)

THE AROMA-CAUSING CHEMICAL CONSTITUENTS OF FLATUS

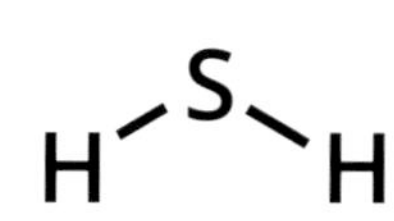

HYDROGEN SULFIDE

smells like sulfur and rotting eggs

CH_3SH

METHANETHIOL

smells like sulfur and garlic

$H_3C{-}S{-}CH_3$

DIMETHYL SULFIDE

smells like cabbage and sulfur

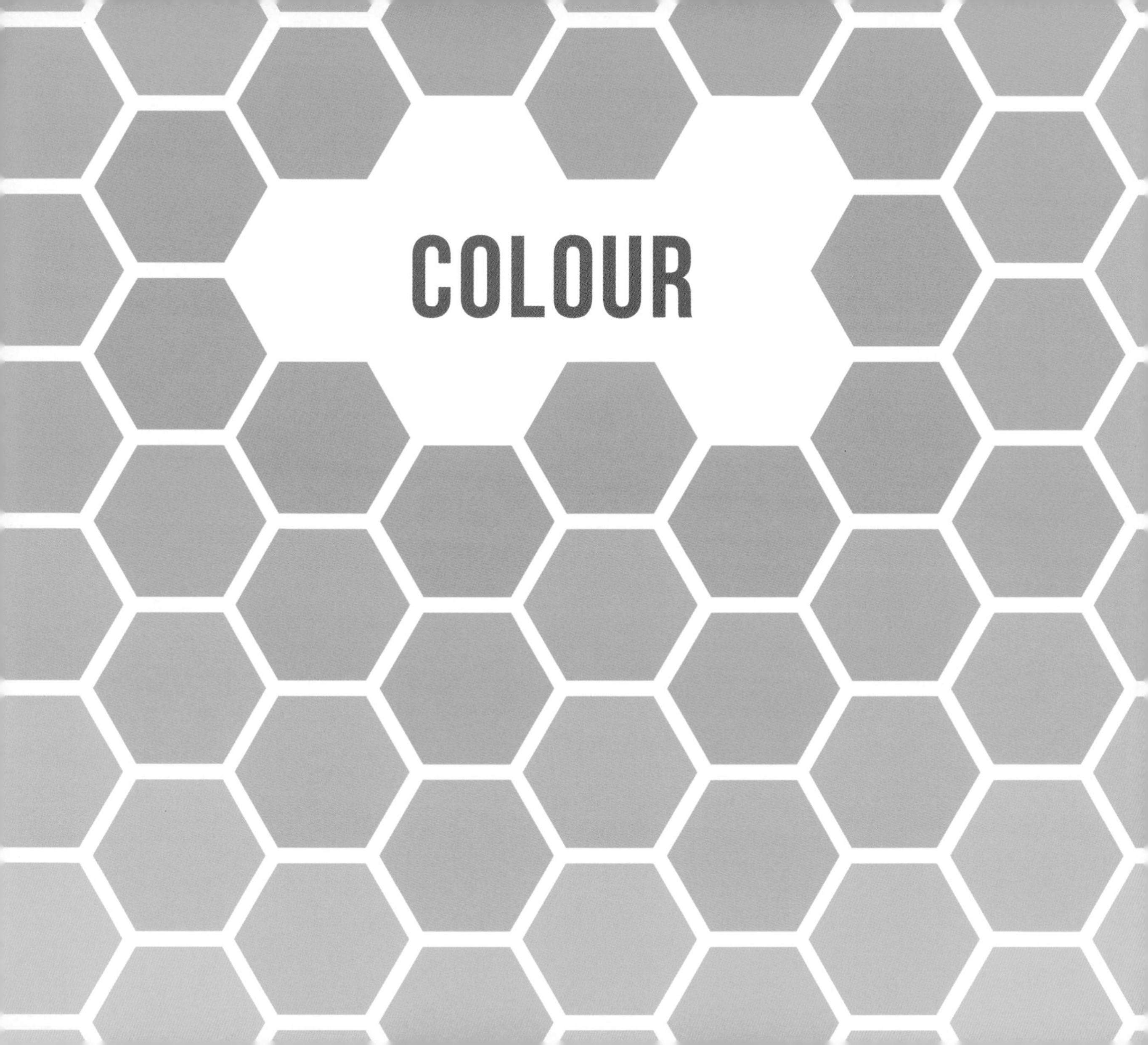

COLOUR

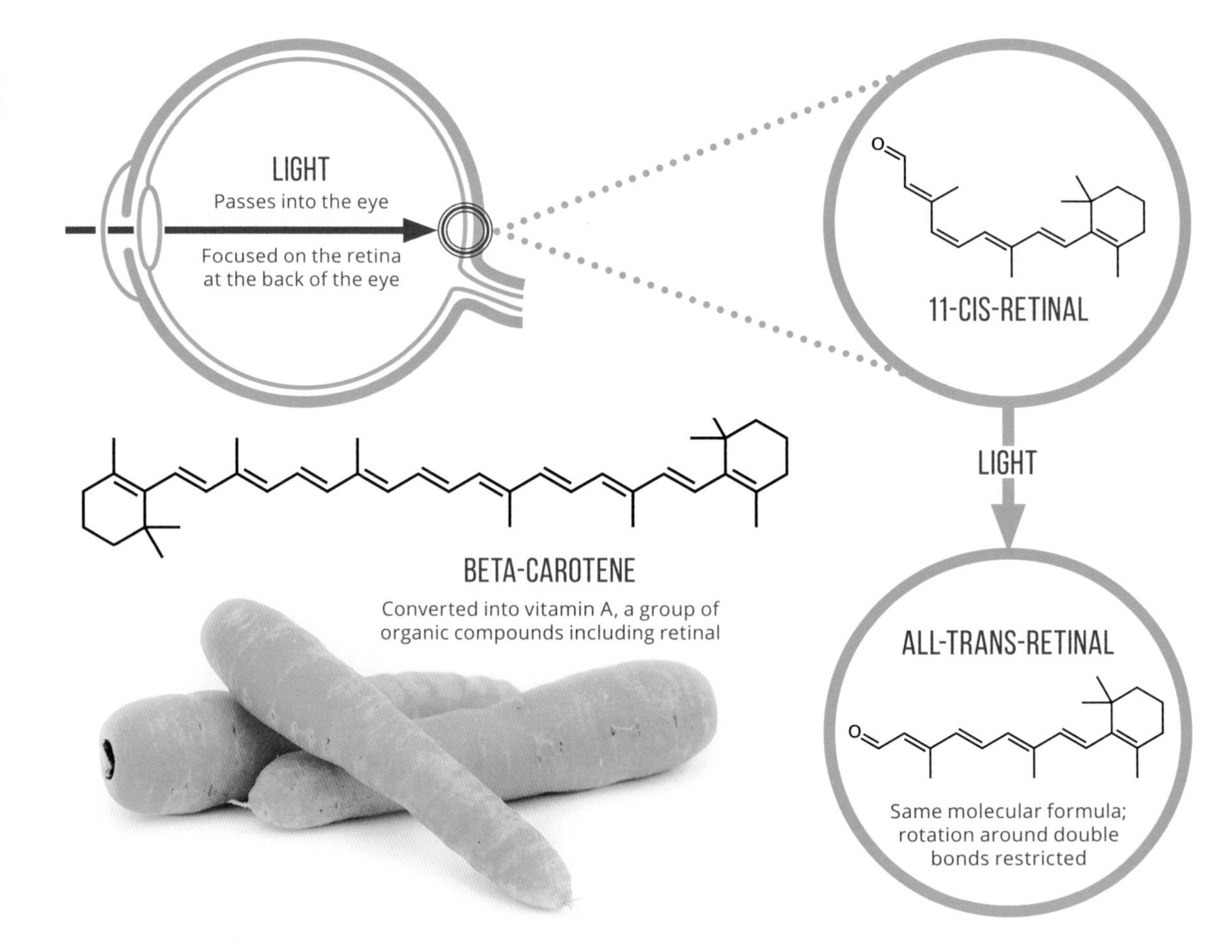
LIGHT
Passes into the eye
Focused on the retina
at the back of the eye
O
11-CIS-RETINAL
LIGHT
BETA-CAROTENE
Converted into vitamin A, a group of
organic compounds including retinal
ALL-TRANS-RETINAL
O
Same molecular formula;
rotation around double
bonds restricted

CAN CARROTS HELP YOU SEE IN THE DARK?

That carrots can help improve your night vision is a commonly espoused claim, and one that's probably responsible for plenty of reluctant children's consumption of the vegetable at the behest of their parents. Whether there's any truth to this claim can be determined by examining the chemical products present, and what happens to them in our bodies.

The orange colour of carrots comes from a chemical they are particularly rich in, beta-carotene. It causes an orange colouration as the bonds in the molecules can absorb specific wavelengths of visible light, resulting in only certain wavelengths being reflected. When ingested, beta-carotene is converted into vitamin A in the liver.

Vitamin A is actually a small group of compounds with very similar chemical structures. They include retinal, the compound that forms the chemical basis for vision in humans and animals. It binds to proteins in the retina of the eye, and also strongly absorbs visible light. The absorption of a photon of light then causes the retinal molecule to convert from one isomer, or form of the compound, to another. When it does this, it can no longer fit into the protein binding site, and effectively shakes itself free. These movements are converted to electrical impulses in the nerve cells in the membrane to which the protein is attached, and these electrical impulses then travel to our brain via the optic nerve, to be interpreted.

On the face of it, then, the claim that carrots can help your vision would seem to have some pretty solid scientific grounding. Retinal is essential for vision, and the beta-carotene in carrots offers a compound from which our body can produce the retinal our eyes require. However, eating carrots will only improve your eyesight if you are vitamin A deficient. This is because the liver simply stores any excess beta-carotene until it's needed, and only a comparatively small amount of vitamin A is needed for vision. A carrot a day actually provides all the beta-carotene your body requires.

It turns out that the idea that carrots can improve your eyesight has its roots in a bit of British propaganda from World War II. After successfully using a new radar system to locate and shoot down German bombers, the British forces came up with the entirely false campaign stating that their pilots were eating carrots to improve their night vision in order to hide the existence of the radar system from the Germans. This campaign of disinformation was so successful that it took root and persists today.

There is one effect that excessive consumption of carrots can have. Eating too many can cause carotene levels in the body to build up to the extent that your skin can develop a yellowy-orange hue. There's even a medical term for this condition: carotenemia.

WHY CAN BEETROOT TURN URINE RED?

An unusual effect of beetroot is that it can cause 'beeturia', or a red colouration to the urine, after ingestion. This condition isn't something that affects just anyone who eats beetroot, so what are the chemical compounds behind it, and why isn't it a universal effect?

Perhaps unsurprisingly, it's the compounds that give beetroot its red colour that can also lead to red-coloured urine. Beetroot's deep red appearance is due to the presence of a class of compounds called betacyanins. This class comprises a number of compounds with similar chemical structures; betanin is a major player as far as the colouration of beetroot goes, and is actually extracted from beetroots and used as a food colouring (named 'Beetroot Red' and designated with the E number E162). Another family of compounds present are the betaxanthins. These have a yellow colour in isolation, and are present in lower concentrations than the betacyanins.

Betacyanins can cause beeturia because they don't break down in the digestive systems of some people. The reasons for this are still a little uncertain; it has been suggested that the compounds are broken down at low stomach acid pH, and that when the stomach acid is not as strong this does not occur. Therefore, the compounds are able to pass through the remainder of the digestive system, absorbed through the intestinal walls in the colon into the bloodstream, then filtered out by the kidneys and into the urine. Of course, some of the unmetabolised compound may well remain in the colon and wind up giving the delightful effect of purple excrement.

It's possible that the breakdown of these compounds could also be influenced by genetic factors which have yet to be precisely determined. For example, if a person genetically has a stronger acidity in their stomach, they may well break the compound down effectively and never experience beeturia. However, studies have shown that there does not appear to be a direct genetic link. In another interesting suggestion, beeturia has been potentially linked to being an early indicator of haemochromatosis (over-accumulation of iron in the body).

Another suggestion of research is that we may all experience beeturia to some extent. One study found that the chemical pigments in beetroot were found in the urine of all of their test subjects, but only in a select number were their concentrations high enough to cause a noticeable colouration.

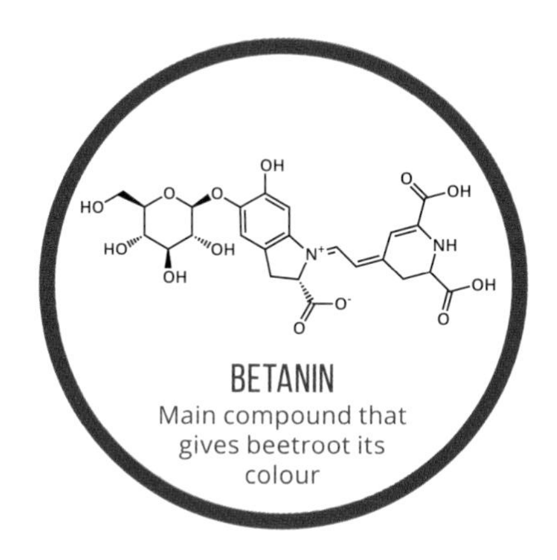

10–14%

OF THE POPULATION ARE AFFECTED

Some studies have suggested that we may all excrete the red pigment to some extent – but only in some people is it concentrated enough to produce a noticeable colouration to urine.

POSSIBLE CAUSES OF BEETURIA

Genetic factors

Iron deficiency

Stomach acidity

LEVELS OF SOLANINE IN POTATOES

Flesh (normal potato)	**~12mg/kg**
Skin (normal potato)	**~150mg/kg**
Skin (green potato)	**~1068mg/kg**
Recommended safe level	**200mg/kg**

SOLANINE
Main poisonous alkaloid

225–450mg Estimated amount of solanine required for a fatal dose (assuming 75kg body weight)

WHY DO POTATOES TURN GREEN?

At some point you've probably left potatoes forgotten somewhere and they've then slowly developed sprouts, or even roots. You may have also noticed that, when left for long enough, either exposed to light or in higher temperatures, they start to take on a light green hue. The general advice is to peel any trace of green away before eating potatoes that have taken on this colouration – and with good reason.

The green colour itself has a harmless enough cause. It's the result of increased production of chlorophyll in the potato, the light-absorbing chemical that allows plants to convert sunlight into energy. However, the colouration that results from this is also thought to be a relatively reliable indicator of the levels of a family of compounds called glycoalkaloids. Glycoalkaloids are a family of toxins that are naturally present in potatoes, even before they turn green, with higher concentrations generally being found in the peel and sprouts of the potato. Usually, their levels are far too low to cause any harmful effects, but the higher levels that can be found in green potatoes, and also in potatoes that have been mechanically damaged, can be enough to elicit symptoms.

The glycoalkaloid family of compounds contains over 90 members, which are found in a number of plants. In potatoes, there are two main compounds, solanine and chaconine, both of which can produce toxic effects in humans. They have fungicidal and insecticidal properties, and are commonly produced by potatoes in response to stress.

Both solanine and chaconine are capable of disrupting cell membranes in the body. The symptoms of glycoalkaloid poisoning include stomach cramps, nausea, diarrhoea and vomiting, and can escalate as far as hallucinations, coma and potentially death. The toxic dose of glycoalkaloids in humans has been estimated to be 2–5 milligrams per kilogram of body weight. So, for an average human being weighing 70 kilograms, 140–350 milligrams would need to be ingested in one sitting. Levels of 200 milligrams per kilogram of potatoes are considered safe. At first glance, this might seem like a risky amount – but you're unlikely to manage a kilogram in one sitting! Also, glycoalkaloid levels in most potatoes are usually much lower than this.

It's worth noting that there is a large variability in the glyco-alkaloid content of potatoes, both due to the range of different varieties, and the fact that those available in stores are developed to contain a lower glycoalkaloid content. So, there's nothing to fear from potatoes before they turn green; those that do turn green can usually be safely eaten if peeled first, though if they're especially bitter tasting this can be an indicator that the flesh is also harbouring a higher concentration. To prevent the potatoes turning green in the first place, it's best to store them somewhere cool, dark and dry.

WHY DO AVOCADOS GO BROWN SO QUICKLY?

Avocados are quick to turn brown after they have ripened. As always, there are chemical processes at work that are to blame for this rather frustrating occurrence.

The flesh of avocados is made up of mainly fatty acids, such as oleic acid and linoleic acid. They contain very little sugar or starch and don't start to ripen until they are picked from the tree.

The rapid browning of avocado flesh is a consequence of its exposure to oxygen in the air, as well as the presence of phenolic compounds in the avocado itself. In the presence of oxygen, an enzyme avocados contain called polyphenol oxidase aids the conversion of phenolic compounds to another class of compounds, quinones. Quinones are capable of polymerising, taking the smaller molecules and joining them together to form a long chain, to produce polymers called polyphenols. This polymerisation manifests itself as a brown colouration to the flesh. The browning doesn't happen in the intact avocado, not only because the flesh isn't exposed to oxygen, but because the phenolic compounds are stored in the vacuole of the plant cells, while the enzymes are found in the surrounding cytoplasm. So, both damage to these cell structures and exposure to oxygen is required for browning to occur.

This browning isn't unique to avocados – the browning of many other fruits, such as apples, is also a consequence of this reaction. For the fruit, it's not a purely aesthetic process. Quinones are compounds that are toxic to bacteria, so their creation from phenolic compounds enables the fruit to last a little longer after exposure to oxygen before beginning to rot.

Browning of avocados can be prevented in several ways. One of the most effective is to rub lemon juice on the exposed flesh of the fruit. The enzymes which enable the enzymatic browning reactions to occur are sensitive to acidic conditions, and work much slower in them. Another option is covering the avocado flesh tightly in cling film. This prevents oxygen from reaching the flesh, and thus browning cannot take place. Chilling the avocado in the fridge can also slow down the enzymes to an extent, as their activity is lower at cooler temperatures. The commonly touted method of leaving the seed pit in the avocado to prevent browning does work – but only on the part of the avocado that it's shielding from oxygen. Exposed areas of the flesh will still turn brown in time.

One final fact about avocados that I feel compelled to include here is to do with the origin of the fruit's name. Whether because of its shape, or because they were thought to consider avocados to have aphrodisiac properties, the Aztecs named the trees it grew on – *ahuacacuahuitl* – which roughly translates as 'testicle tree' and 'guacamole' derives from the Aztec word *ahuacamolli*, which translates as testicle soup. Lovely. It seems unclear as to which came first in the Aztec lexicon, and it's entirely possible that their word for avocado was used euphemistically for 'testicle', rather than the other way around. Either way, next time you're eating avocados or guacamole, it's a great fact to unsettle your fellow diners with.

0 HOURS

3 HOURS

6 HOURS

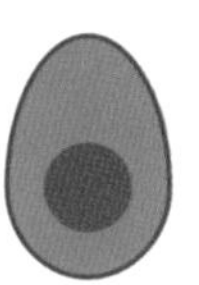
9 HOURS

12 HOURS

ENZYME
(polyphenol oxidase)
→
OXYGEN

CATECHOL
(type of polyphenol)

1,2-BENZOQUINONE

A MELANIN

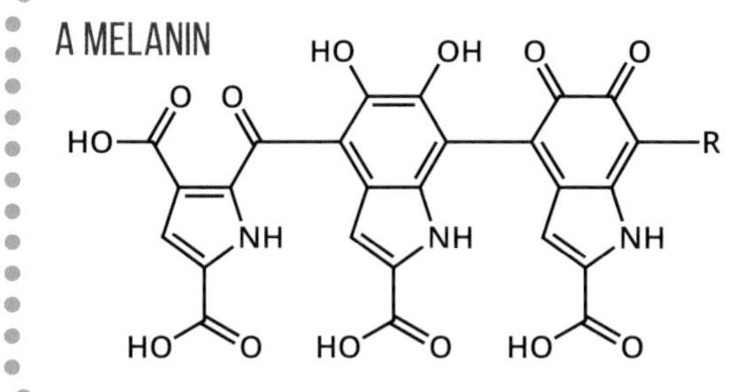

Polymeric pigments called melanins cause the brown colouration. Melanins are also the primary pigments determining skin colour in humans.

DELAYING AVOCADOS TURNING BROWN

Cover with cling film

Add lemon juice

Store in fridge

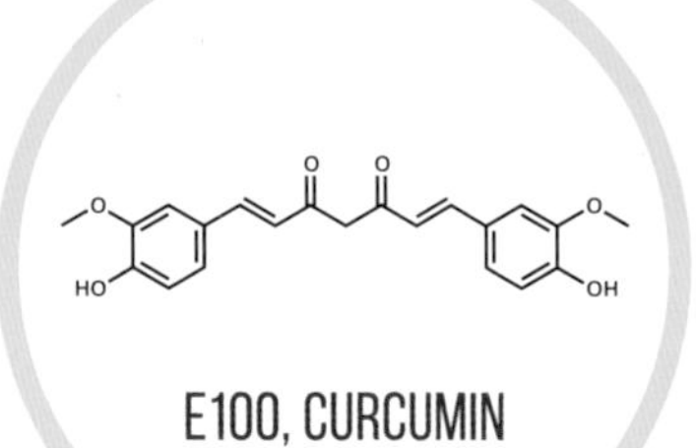

E100, CURCUMIN
natural yellow 3

WHAT ARE E NUMBERS?

Substances that are allowed to be used as food additives within the EU are categorised using E numbers. The numbers E100–199 are used for food colourings. There are also E numbers for preservatives, flavour enhancers, sweeteners and thickeners, as well as other chemical agents commonly added to foods.

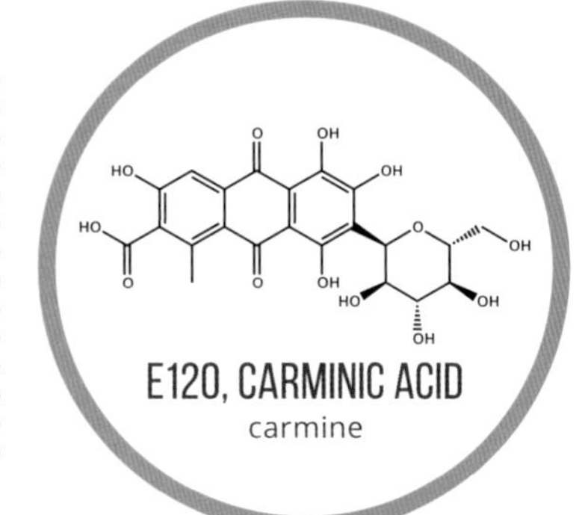

E120, CARMINIC ACID
carmine

E132, INDIGO CARMINE
indigotine

E160E, APOCAROTENAL
food orange 6

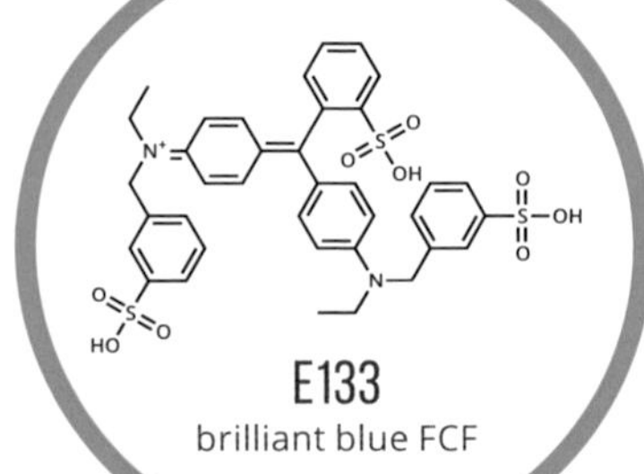

E133
brilliant blue FCF

WHAT CAUSES THE DIFFERENT COLOURS OF FOOD COLOURINGS?

There are a wide range of food colourings, both synthetic and natural, that can be added to foods. A huge number of foods we eat contain them – some of them are even present naturally in some foodstuffs. What is it, though, that causes them to be coloured?

If you look at some of the molecules in the graphic, you'll notice that their structures all have one thing in common: in some part of the structure, they all contain alternating double and single bonds between carbons. In chemistry, this is known as 'conjugation'. This means that the electrons in these bonds, rather than being rigidly fixed in the double bond and single bond areas, are averaged out across the alternating double and single bonds.

That's all well and good, but how does it affect the molecule's colour? Well, electrons in molecules can absorb light. They start off in what we call the 'ground state' – when they absorb energy we refer to them as being in an 'excited state'. In these conjugated molecules, the energy required to get from the ground state to the excited state is at an energy that corresponds to wavelengths in visible light. When visible light hits these molecules, these specific wavelengths of light are absorbed by the electrons, and the rest of the light passes through. The colour we see therefore depends on what colours of light were removed by the absorption of the molecule.

In similar molecules, we can hazard a guess at the colour they would be based on the number of alternating double and single bonds. Generally, though, it's quite hard to predict, due to the fact that different molecules can often have completely different structures, or different functional groups.

Food colourings have been dogged with claims of effects on health for a fair few years and this is justified in some cases. Some synthetic food dyes, which have since been banned from use in foods, have been linked with cancer at high doses. Six naturally derived food colourings have also been voluntarily phased out after studies showed they have the potential to cause hyperactivity in children; these colourings are commonly referred to as the 'Southampton Six', as the study was carried out at Southampton University. Any products that still use one of these six dyes must display the fact prominently on the packaging. All the other food dyes, however, have passed stringent safety tests and are regarded as harmless when added to foods.

WHY ARE SALMON AND PRAWNS PINK?

Salmon is a bit of an oddity when you compare it to the pasty white colour of the other fish we regularly eat, such as cod and haddock. The cause of its pink colour, as well as the pink hue that crustaceans such as prawns take on when cooked, is down to a particular chemical compound, astaxanthin.

Astaxanthin is a member of the carotenoid family of compounds; this family has over 600 members, and they are commonly found as pigments in plants. In the wild, algae and other micro-organisms produce astaxanthin as a by-product, which is then consumed by other sea creatures including prawns and small fish. When salmon eat these creatures, they also take in the astaxanthin they contain. As it can build up in their flesh, it leads to the familiar pink colour that we associate with salmon.

Obviously, many of the salmon we eat today are farmed salmon, which can impact on the amount of astaxanthin that they accumulate and lead to their flesh not looking very pink at all. To remedy this, synthetic astaxanthin, as well as astaxanthin from biological sources, can be added to their feed to ensure their colour is still as expected by the time they reach the supermarket. There is no danger in this method of introducing colour, as both synthetic and natural astaxanthin are one and the same compound.

As already mentioned, prawns also contain astaxanthin – so why is it that they only turn pink when cooked? This is because in the raw prawns (and other crustaceans) it is held in a complex with proteins called crustacyanins, which modify the light it absorbs, and therefore give the blue-grey colour of the flesh prior to cooking. When exposed to heat, the crustacyanins are broken down, leading to the usual pink colour of astaxanthin becoming visible. The reason that salmon flesh doesn't need cooking to exhibit the pink colour is because the salmon breaks down the proteins during digestion.

Astaxanthin is also the reason that flamingos are pink. Their diet of prawns and shrimp allows them to accumulate astaxanthin in a similar manner to salmon, and contributes the pink hue to their plumage.

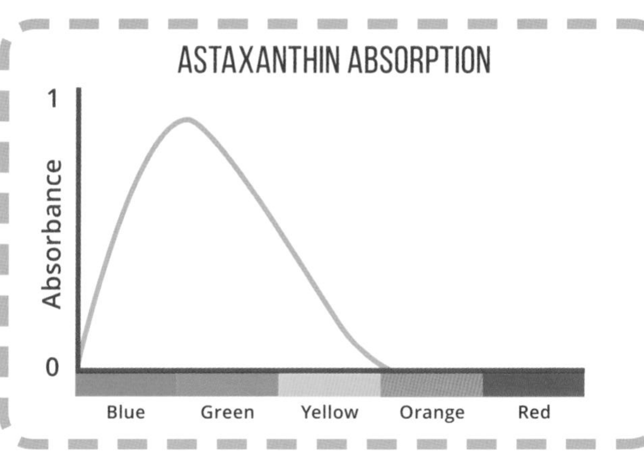

ASTAXANTHIN – CONJUGATION & COLOUR

O

HO

OH

O

Alternating double and single bonds in a molecule are referred to as 'conjugation'.

Highly conjugated molecules (those with lots of alternating double and single bonds) absorb visible light and appear coloured.

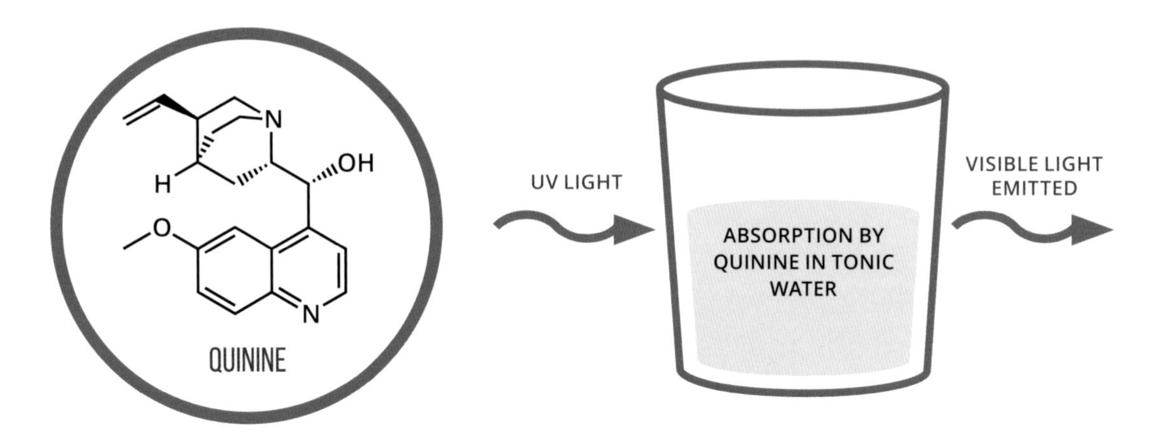

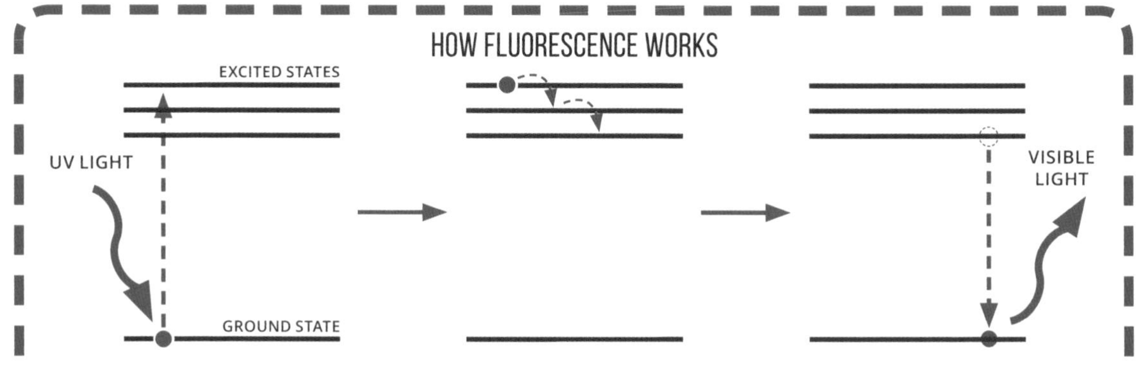

Electrons in the molecule absorb energy from the UV light, and go from their starting energy (the ground state) to a higher energy (an excited state). This is unstable, and they eventually fall back to the ground state, emitting their excess energy as visible light.

WHY DOES TONIC WATER GLOW UNDER A BLACK LIGHT?

If you're a gin and tonic drinker, or even if you're not, here's an experiment to try the next time you happen to have a black light handy. If you shine it on a glass or bottle of tonic water in a dark room, you can observe a weird phenomenon: the tonic water will glow a bright blue. This can't be observed with normal tap water, so what does the tonic water contain that makes it glow so brightly?

The glow is down to the presence of one of the chemical constituents of tonic water, a compound called quinine. Quinine's molecules are capable of absorbing wavelengths of ultraviolet light from the black light – in particular, light with a wavelength of 250 nanometres and 350 nanometres. This absorption of light photons 'excites' electrons in the molecule to higher levels of energy than they would usually occupy. The electron does not remain in this higher-level state for long, however; it gradually falls back down to its original energy level, and as it does so it will emit its excess energy as light. This emitted light is at a longer wavelength than the light initially absorbed, meaning it is in the visible blue region of the spectrum rather than the invisible UV region. As a result of this process, termed fluorescence, we see the tonic water glowing a bright light blue colour.

We know that it's the quinine present in tonic water that's responsible for this effect, as if you try it with quinine-free tonic water, you won't observe any fluorescence. However, quinine isn't the only molecule capable of exhibiting fluorescence – there are many other natural examples. A particularly interesting one is found in the species of jellyfish, *Aequorea victoria*, also called the 'crystal jelly'. It contains a protein which when it reacts with calcium ions in the body of the jellyfish produces blue light. This blue light is converted to green light by another protein, via fluorescence, causing a green glow.

Fluorescence isn't limited to nature, either. Bank notes can often have fluorescent security features built in, to make them harder to counterfeit. As an example, English bank notes all feature fluorescent dyes, which cause the numbers of the value of the note to fluoresce red and green when they are placed under a UV light.

POISON

HAEMAGGLUTINATION UNITS (HAU)

70,000	200–400
UNCOOKED	COOKED

The protein, phytohaemagglutinin, is responsible for the poisonous nature of raw kidney beans.

Canned kidney beans in shops are safe to eat from the can, as they are specially treated before packaging:

SOAKED FOR SEVERAL HOURS & **BOILED** FOR THIRTY MINUTES.

HAEMAGGLUTINATING UNIT CONTENT OF OTHER BEAN VARIETIES

CANNELLINI BEANS	BROAD BEANS
~30%	~5–10%

Percentages are for raw beans, compared to haemagglutinating units in raw kidney beans.

4–5 RAW KIDNEY BEANS

This quantity of raw beans is enough to induce the symptoms of poisoning within 3 hours, which include:

- NAUSEA
- VOMITING
- DIARRHOEA
- ABDOMINAL PAIN

WHY ARE KIDNEY BEANS POISONOUS IF UNCOOKED?

Kidney beans are an essential ingredient in a good chilli con carne. However, they also have a more sinister side, in that, unless they are cooked, they harbour a potent toxin that can lead to illness, or even death in extreme cases. Luckily, there are measures taken to ensure that the canned kidney beans bought in supermarkets are already safe for consumption, but this toxicity is still a very real risk with raw kidney beans.

The toxin responsible for the potentially poisonous nature of kidney beans is actually a protein, called phytohaemagglutinin, often abbreviated to PHA. This protein is present in the beans because it helps protect them against pests and pathogens. In humans, however, PHA can cause duplication of cells (mitosis), affect cell membranes, and cause red blood cells to clump together.

PHA is actually found in many other types of beans, although in much lower concentrations than in red and white kidney beans. Its presence is usually measured in haemagglutinating units, hau. When raw, red kidney beans can contain up to 70,000 hau; by comparison, uncooked broad beans have only 5–10 per cent of this amount.

Eating as few as four or five raw red kidney beans can be enough to initiate symptoms of poisoning, which include nausea, vomiting and diarrhoea. Eating a large enough number could potentially be deadly, though recorded cases are rare. In most cases of poisoning, unless a large number of beans have been ingested, hospital treatment is not required and the symptoms subside after several hours.

Why is it, then, that we have nothing to fear from the canned kidney beans available in supermarkets? The reason is that these beans are specifically prepared in order to minimise the concentration of the toxin before they make it to the shelves. They will be soaked for a period of several hours and then boiled for thirty minutes. The heat breaks down the toxin, and makes the kidney beans safe to eat; compared to the high levels beforehand, cooked kidney beans only contain about 200–400 hau of PHA. Some of the instances of kidney bean poisoning have come from people using raw kidney beans and cooking them in a slow cooker, which doesn't reach a high enough temperature to break down PHA. In fact, it can increase the concentration of the protein, so it's important to ensure that, if you do cook with raw kidney beans, they are cooked through thoroughly.

WHY ARE SOME MUSHROOMS POISONOUS?

There's a reason that it's strongly recommended not to pick wild mushrooms unless you've had training in recognising the different types; some mushrooms containing deadly toxins can look almost identical to those that are perfectly safe to eat. Of the various types of mushroom toxins, those which cause the greatest number of deaths are the amatoxins and orellanine.

The sinisterly named 'death cap' and 'destroying angel' mushrooms both contain amatoxins. The amatoxins are a family of structurally similar compounds, with minor changes in parts of the structure determining the different types, of which ten are currently known. The main amatoxins commonly found in significant quantities are α-amanitin, β-amanitin and γ-amanitin, all three of which have a median lethal dose of around 0.5–0.75 milligrams per kilogram of body weight.

It can take between 6 and 24 hours for the symptoms of amatoxin poisoning to begin to manifest. The initial symptoms are stomach cramps, vomiting and diarrhoea; these can actually improve after a few days, but ultimately the toxin can cause liver and kidney failure, leading to death within five to eight days of consumption of the mushrooms. It's estimated that between 10–20 per cent of diagnosed cases of amatoxin poisoning result in death, with many of those that survive requiring liver transplants to do so.

The deadly webcap and fool's webcap mushrooms contain orellanine; this particular toxin initially causes thirst, stomach cramps and nausea, and can go on to cause a low output (or even no output) of urine. The initial symptoms can take up to three weeks to appear, though usually they are notable two to three days after ingestion. The later symptoms are due to kidney damage, which can, in severe cases, culminate in kidney failure. Again, in these cases, transplant is often the only option to treat the poisoning, with no known antidote for orellanine.

The most recognisable poisonous mushroom is probably fly agaric. This red, white-spotted specimen contains the compound muscarine, although in lower concentrations than some other mushroom species – it's estimated that it only constitutes around 0.0003 per cent of the mushroom's weight. Muscarine was originally thought to be the source of the toxicity of fly agaric, but it has since been discovered that another compound, muscimol, is largely responsible. It's also found in another common poisonous mushroom, the panther cap. No deaths have been officially attributed to either fly agaric or panther caps, but their ingestion can cause dizziness, stomach irritation and hallucinogenic effects.

Unfortunately, there's no telltale clue when it comes to spotting which mushrooms are poisonous and which are not. Some of the deadliest can taste delicious, and look benign.

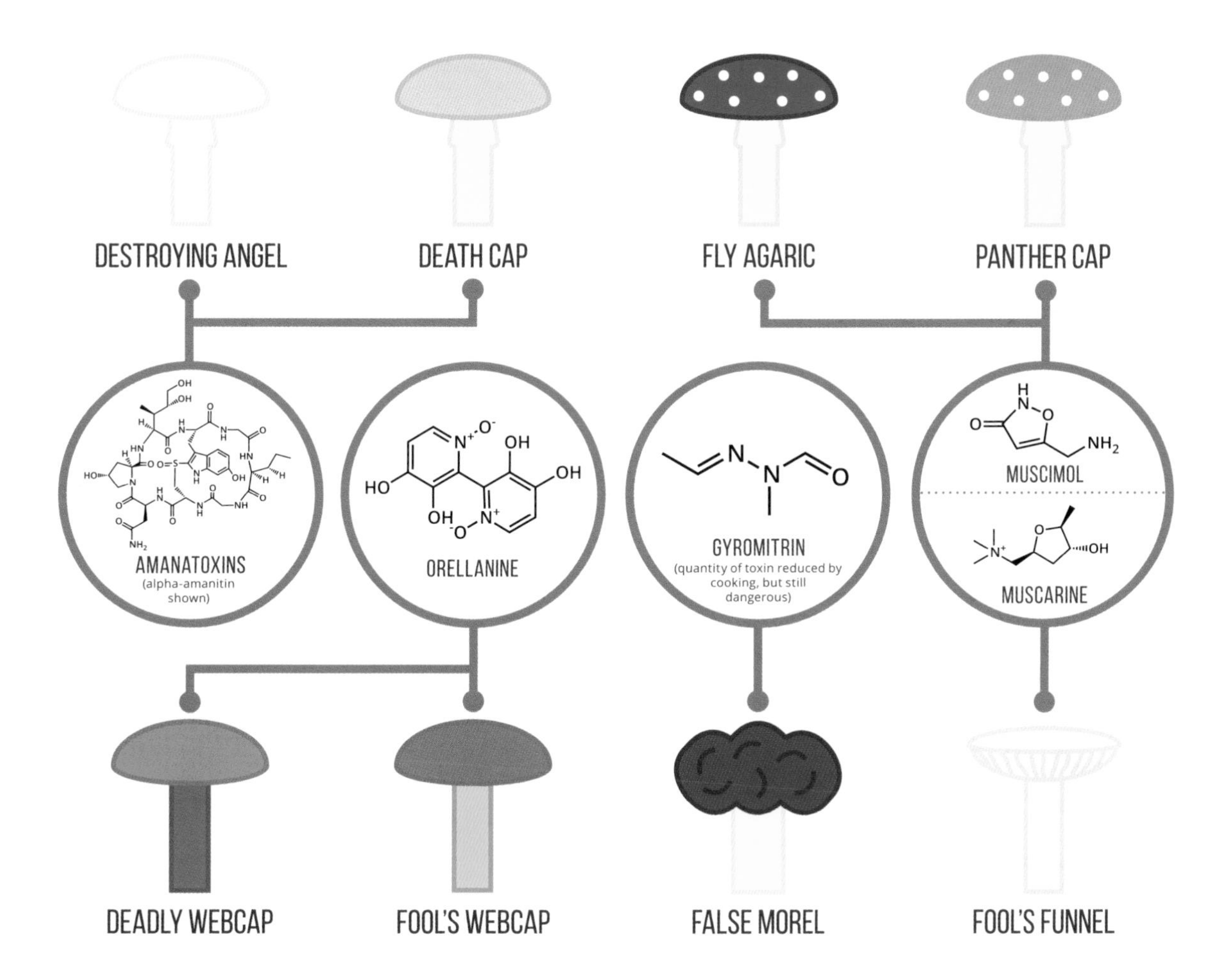
DESTROYING ANGEL
DEATH CAP
FLY AGARIC
PANTHER CAP
AMANATOXINS
(alpha-amanitin shown)
ORELLANINE
GYROMITRIN
(quantity of toxin reduced by cooking, but still dangerous)
MUSCIMOL
MUSCARINE
DEADLY WEBCAP
FOOL'S WEBCAP
FALSE MOREL
FOOL'S FUNNEL

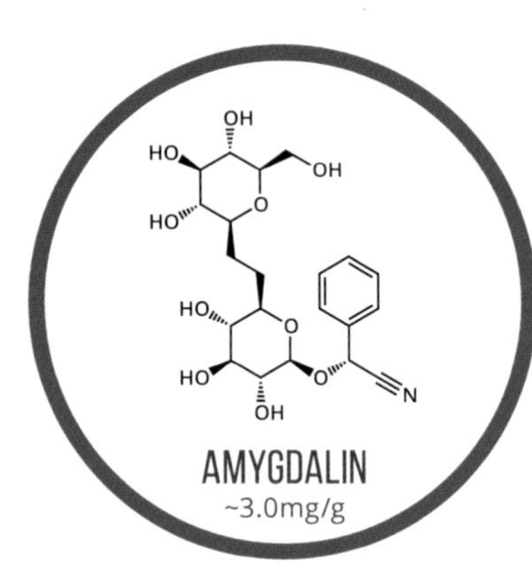

AMYGDALIN
~3.0mg/g

H—C≡N

HYDROGEN CYANIDE

A highly poisonous gas, produced by metabolism of amygdalin in the body

THE SEEDS OF MANY OTHER FRUITS ALSO CONTAIN VARYING QUANTITIES OF AMYGDALIN

APRICOT
~14.4mg/g

BLACK CHERRY
~2.7mg/g

RED CHERRY
~3.9mg/g

NECTARINE
~0.1mg/g

PLUM
~2.2mg/g

PEACH
~6.8mg/g

PEAR
~1.3mg/g

GREENGAGE
~17.5mg/g

Provided values refer to estimated average amygdalin content of the seeds of the fruit.

AMYGDALIN LD_{50}

675–3,750mg

The reported range for a lethal dose of amygdalin in a human of average weight (75kg), in the absence of any physical or medical conditions.

DO APPLE SEEDS REALLY CONTAIN CYANIDE?

It's commonly mentioned that apple seeds contain cyanide, and that you shouldn't eat them to avoid the risk of inadvertently poisoning yourself. Now, this might sound somewhat far-fetched, but there's actually more than a grain of truth to it.

First and foremost, you're highly unlikely to manage to eat enough apple seeds in order to poison yourself, so you can rest easy if you occasionally accidentally swallow one. Apples actually contain a compound called amygdalin in their seeds, which is a cyanide-and-sugar-based molecule. While not toxic when the seeds are intact, if the seed is chewed or otherwise broken, human or animal enzymes come into contact with the amygdalin and effectively cut off the sugar part of the molecule. The remainder of the molecule can then decompose to produce the poisonous gas hydrogen cyanide. But this only happens if the seed has been somehow damaged – otherwise, it will pass through unaffected.

Cyanide toxicity is experienced by humans at doses of around 0.5–3.5 milligrams per kilogram of body weight. Symptoms of cyanide poisoning include stomach cramps, headache, nausea and vomiting, and can culminate in cardiac arrest, respiratory failure, coma and death. A fatal dose for humans can be as low as 1.5 milligrams per kilogram of body weight. In the form of Zyklon B, hydrogen cyanide was the chemical entity used by the Nazis in their concentration camps in World War II.

In a recent study, the amygdalin content of apple seeds was found to be approximately 3 milligrams per gram of seeds (one seed is approximately 0.7g). As not all of this mass would be converted into hydrogen cyanide (some of it will constitute the sugar part of the molecules that is cleaved off), it's apparent that you're going to need to eat a huge number of apple seeds to succeed in poisoning yourself, and there don't appear to be any cases of someone having succeeded in doing so.

It's not just apples that contain cyanide-releasing compounds. Many other fruit seeds contain amygdalin, including cherries, peaches, plums and pears. The much larger apricot kernels can contain up to almost 15 milligrams of amygdalin per gram of seeds, and in their case there are recorded instances of hospitalisation. A research paper in 1998 reported the case of a 41-year-old woman who was admitted to hospital with cyanide poisoning after eating approximately 30 bitter apricot kernels (she was eating them as a health food and presumably got carried away). This could potentially have been fatal if not for the administering of antidotal treatment; as it was, she went on to make a full recovery.

WHAT CAUSES SHELLFISH POISONING?

It's common knowledge that eating a bad oyster or mussel during a shellfish dinner can result in some pretty unpleasant side effects. What's perhaps not as well known is that there's not one, but four different types of shellfish poisoning you can potentially contract: diarrhetic, neurotoxic, amnesic and paralytic. Each of these are caused by differing chemical compounds, and have different symptoms.

Shellfish are filter feeders, which means they feed by drawing water into their shells and straining out food particles. It's estimated that oysters filter around five litres of water per hour. In particular, their diet consists of dinoflagellates, a type of single-celled organism, and marine plankton. Large concentrations of these organisms in seawater can lead to an effect known as a 'red tide' or 'algal bloom'. These blooms can be harmful to marine life, and can eventually lead to shellfish poisoning, as shellfish feed on them and accumulate high concentrations of the toxins produced.

Perhaps the most benign of the four poisoning types is diarrhetic poisoning. This particular brand manifests itself mainly through the incidence of diarrhoea, though nausea, stomach cramps and vomiting are also common symptoms. A particular type of dinoflagellate is responsible for the production of okadaic acid, the chemical culprit causing diarrhetic poisoning. As little as 48 micrograms (0.000048 grams) is required in order to kick-start symptoms, which can last as long as three days, and it isn't broken down at all by cooking. It exerts its effect by causing cells to become very water permeable, impairing the water balance of the intestines and leading to diarrhoea. Though uncomfortable, there are no recorded deaths from diarrhetic poisoning.

You're likely to survive neurotoxic poisoning too, though there have been recorded incidences of hospitalisation. In this case, a whole class of ten or more compounds, known as brevetoxins, are responsible for the symptoms. These include nausea, vomiting, slurred speech, a prickly feeling in the mouth, lips or tongue, and in rare cases even temporary partial paralysis. Again, these compounds aren't broken down by heat, so they're still a risk even after cooking.

The third type of poisoning, amnesic poisoning, is caused by the production of the toxin domoic acid by a type of algae known as diatoms. Symptoms it induces consist of nausea, vomiting and diarrhoea, as well neurological symptoms including headaches, short-term memory loss, and in severe cases seizures, heart arrhythmia and even death. Once again, domoic acid is heat stable, and there is no known antidote.

Paralytic poisoning, the last of the four types, is potentially the most life-threatening. It's caused primarily by saxitoxin, produced by dinoflagellates and cyanobacteria in sea water. Saxitoxin interferes with the transmission of signals from nerve cells, which can lead to paralysis. As well as the common symptoms of nausea, vomiting and diarrhoea, the slow paralysis of muscle tissue can lead to cardiac arrest and respiratory failure.

Due to the potentially deadly effects of shellfish poisoning, shellfish harvesting can often be suspended in the advent of red tides in the area in order to prevent the possibility of poisoning cases. Toxic shellfish will look and taste no different to normal shellfish, however, so if you do experience the symptoms of shellfish poisoning after a meal, it's always recommended to consult a doctor immediately.

Diarrhetic poisoning

OKADAIC ACID

diarrhoea, nausea, vomiting, stomach cramps

Neurotoxic poisoning

BREVETOXINS

nausea, vomiting, slurred speech, prickliness, partial paralysis (rare)

Amnesic poisoning

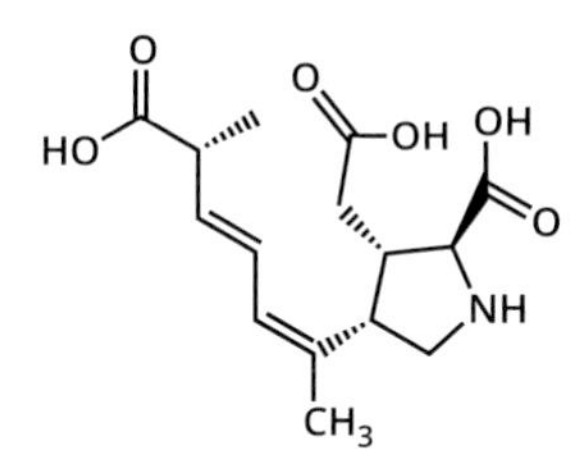

DOMOIC ACID

nausea, vomiting, diarrhoea, headaches, short-term memory loss, seizures (severe cases)

Paralytic poisoning

SAXITOXIN

nausea, vomiting, diarrhoea, paralysis, cardiac arrest, respiratory failure

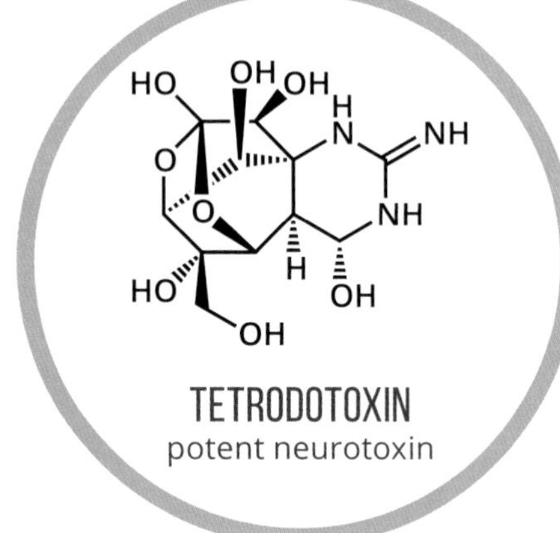

TETRODOTOXIN
potent neurotoxin

HOW DOES TETRODOTOXIN AFFECT THE BODY?

Prevents the normal signalling between the body and brain

This leads to the paralysis of voluntary muscles in the body

Diaphragm and intercostal muscles are affected, stopping breathing

Heart rate becomes unregulated and can rise to 100bpm

ESTIMATED HUMAN MEDIAN LETHAL DOSE OF TETRODOTOXIN
0.33 MG/KG OF BODY WEIGHT

HARMFUL LEVELS OF TETRODOTOXIN ARE FOUND IN THE
LIVER, SKIN & GONADS OF THE PUFFERFISH

WHY IS EATING PUFFERFISH A RISKY MOVE?

While not a common food in the Western world, pufferfish is a notorious dish in Japan, where dishes prepared from it are known as fugu. The appeal of the dish for many is the sense of dicing with death, for pufferfish contain a toxin that can prove deadly in even tiny amounts: tetrodotoxin.

Tetrodotoxin is actually produced by bacteria that reside within the pufferfish. It's also not limited to pufferfish, with the blue-ringed octopus and several other fish species also harbouring it, but pufferfish is the most commonly eaten of those in which it occurs. While pufferfish has been eaten for centuries in Japan, its preparation is very important. In particular, the liver, gonads and skin of the fish contain high levels of tetrodotoxin, and these must be skilfully removed by the chef before the fish is served as food. Even cooking the fish doesn't remove the toxin, as it's heat stable and doesn't break down.

Although the pufferfish itself is not susceptible to its own poison, humans are very much at risk. It's been estimated that tetrodotoxin is approximately 1,200 times as poisonous as cyanide, and a rough median lethal dose has been estimated at approximately 25 milligrams for an average person weighing 75 kilograms. The toxin acts quickly if ingested; it blocks sodium channels, which are important in our bodies for the relaying of messages between the brain and the body. As a result, it induces paralysis of muscles – victims remain fully conscious as the paralysis worsens, until eventually the muscles which enable breathing are also paralysed and asphyxiation occurs.

There is no known antidote to tetrodotoxin. Chefs in Japan must earn a licence that allows them to prepare and serve fugu, so that diners can be assured that the poisonous parts of the fish have been removed. The licence requires three years of training, and only around 35 per cent of applicants pass the eventual test and receive the licence. There are still a handful of cases of tetrodotoxin poisonings a year, but many of these deaths are the consequence of fishermen trying to prepare pufferfish themselves.

In recent years, scientists have managed to breed a variety of pufferfish that does not contain the toxin, and therefore does not require such careful preparation. However, many fugu chefs have suggested that they would continue to use the variety of the fish containing the toxin – they suspect that the appeal of the dish comes primarily from the concept of dicing with death when eating it.

WHY IS CHOCOLATE POISONOUS TO DOGS?

Theobromine is a stimulant compound found in chocolate; it comes from the same family of compounds as caffeine, and its chemical structure is very similar. It also acts on the body in a similar manner to caffeine, blocking certain receptors in the brain and reducing sleepiness. All types of chocolate contain theobromine, though it is highest in dark chocolate, while white chocolate only contains minute trace amounts.

Theobromine is the culprit when it comes to chocolate's toxicity. In humans, the median lethal dose (the dose required to kill 50 per cent of a test population) is unknown, on account of no one ever having eaten enough chocolate in order to induce death from theobromine poisoning. Estimates for the exact figure vary significantly, but it's suggested that you'd need to eat 5 kilograms of milk chocolate at the very least in order to ingest anywhere near the amount of theobromine required for poisoning.

If you compare this to the median lethal dose for dogs, it's easy to see why chocolate is far more toxic for canines. The figure stands at 300 milligrams of theobromine per kilogram of weight; assuming we're talking a relatively small dog of around 10 kilograms, it'd need to consume 3 grams of theobromine to reach this dose, which equates to roughly 2 kilograms of milk chocolate. This figure still seems quite large, but dark chocolate's theobromine content can be as high as 600 milligrams per 100 grams, so the chocolate required to reach the dose drops to around 500 grams. It's also worth mentioning that the symptoms of theobromine poisoning, which include vomiting and diarrhoea, would certainly kick in well before this.

You might be wondering why it is that theobromine has such a potent effect on dogs in comparison to humans. The reason behind this is that dogs metabolise, or break down, the theobromine much slower than humans. As a consequence the amount required to cause poisoning can quickly build up. Cats actually have an even lower tolerance for theobromine, but as they lack the ability to taste sweetness, they're less prone to start eating any chocolate that's left lying around.

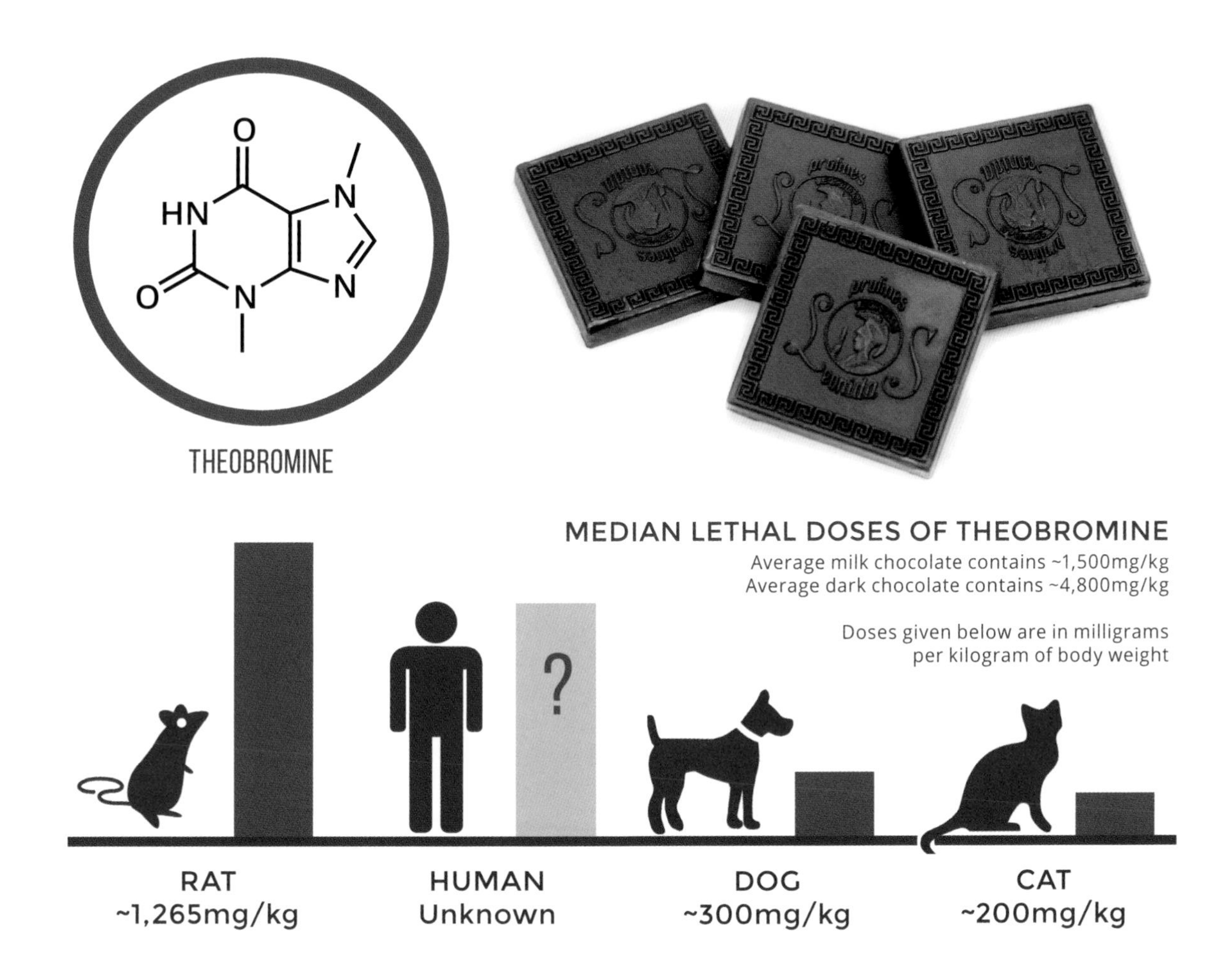
O
HN
N
O
N
N
THEOBROMINE
MEDIAN LETHAL DOSES OF THEOBROMINE
Average milk chocolate contains ~1,500mg/kg
Average dark chocolate contains ~4,800mg/kg
Doses given below are in milligrams
per kilogram of body weight
?
RAT
~1,265mg/kg
HUMAN
Unknown
DOG
~300mg/kg
CAT
~200mg/kg

CAN MIXING DRINKS WORSEN YOUR HANGOVER?

As we all know, a night of drinking doesn't always come without a price – drink too much, a splitting headache, lethargy, nausea and perhaps even getting up close and personal with the toilet bowl awaits. Some hangovers always seem to be exponentially worse than others, though, particularly when you've been drinking all sorts of alcoholic drinks the night before. The idea of not mixing drinks is oft quoted, but is there any truth to the idea that it can worsen your hangover?

First, we should consider what's actually going on in your body after you drink alcohol. This process happens the same way, regardless of the type of alcoholic drink you've been drinking, because the alcohol content is always the same chemical compound: ethanol. Up to a maximum of 8 per cent of ethanol you ingest is disposed of by your body through your breath, sweat or urine – hence why a heavy session of drinking can leave you literally stinking of alcohol. The remainder of the alcohol is broken down in your body into other products, and this is what leads to the symptoms of a hangover.

The majority of ethanol is broken down in the liver, where it is converted into acetaldehyde. In turn, this is broken down to acetic acid, which is then converted to a compound called acetyl coenzyme A.

Acetaldehyde is one of the potential key players in hangovers. Your liver is well equipped to cope with small amounts of it, but its store of glutathione, essential for breaking down acetaldehyde, is limited. Drinking large amounts of alcohol can cause this store to run out, and then your body has to wait for more glutathione to be produced before it can break down further amounts of acetaldehyde. In the meantime, if you continue drinking, acetaldehyde levels will build up as your body waits to metabolise it. It's a toxic compound, and its build up can cause headaches, nausea, vomiting and sensitivity to light and sound – symptoms that are uncannily similar to those of a hangover.

Different drinks are also suggested to have an effect as they can have different levels of congeners. These congeners are other, minor, compounds produced during the fermentation process that results in ethanol. They can include other types of alcohol, such as methanol, as well as a range of other organic compounds. Whisky, wine, tequila and brandy all have higher levels of congeners than clear spirits such as vodka and gin, and it's suggested that these may contribute to worse hangover symptoms. A 2009 study which examined the effect of high and low congener beverages on hangovers found that the subjects drinking bourbon consistently rated their hangovers as worse than the subjects drinking the lower congener vodka.

So, does mixing drinks worsen your hangover? While, surprisingly, we still don't know the full picture behind how hangovers occur, it seems clear that the primary cause of a hangover is simply the amount of alcohol you drank the night before. The mixing of drinks itself doesn't seem to have any effect on the severity of the hangover, so we can chalk this one up as a myth.

METABOLISM OF ALCOHOL IN THE BODY ENZYMES METABOLISE ETHANOL AT APPROXIMATELY 10 GRAMS PER HOUR

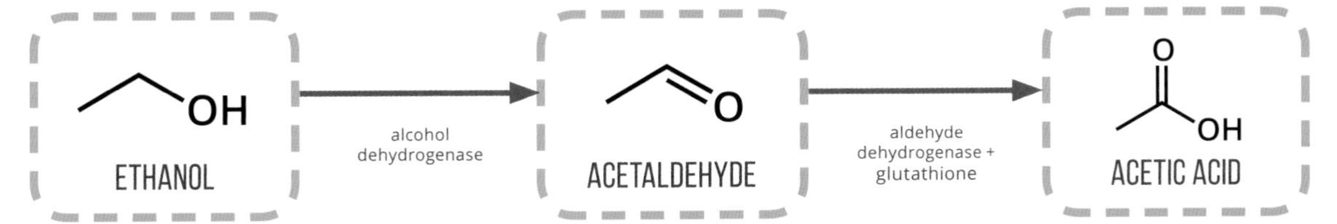

THE SYMPTOMS OF A HANGOVER

TIREDNESS

HEADACHE

LIGHT SENSITIVITY

THIRST

NAUSEA

Other symptoms include muscle aches, shaking and tremors, and, occasionally, vomiting and diarrhoea.

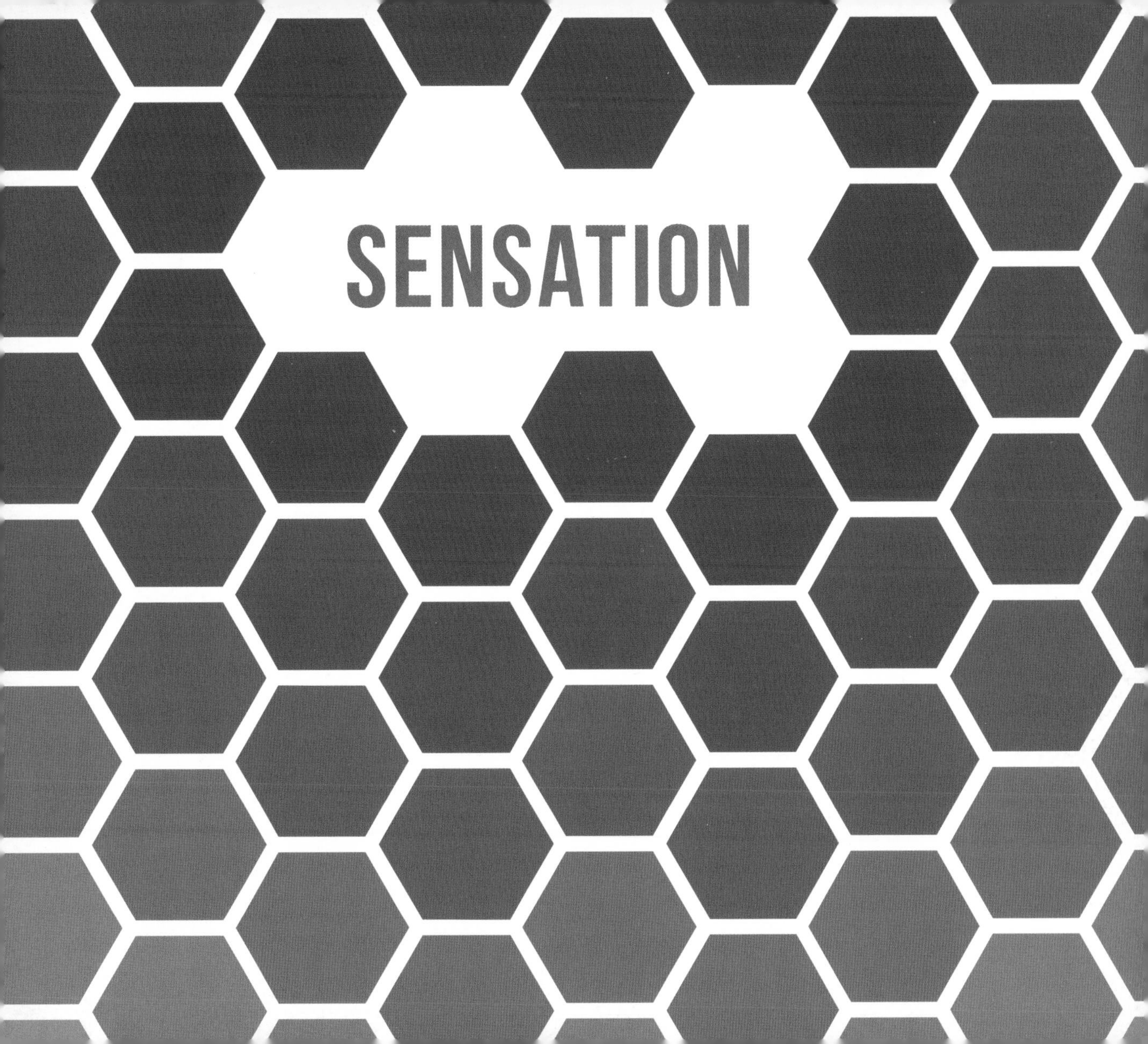
SENSATION

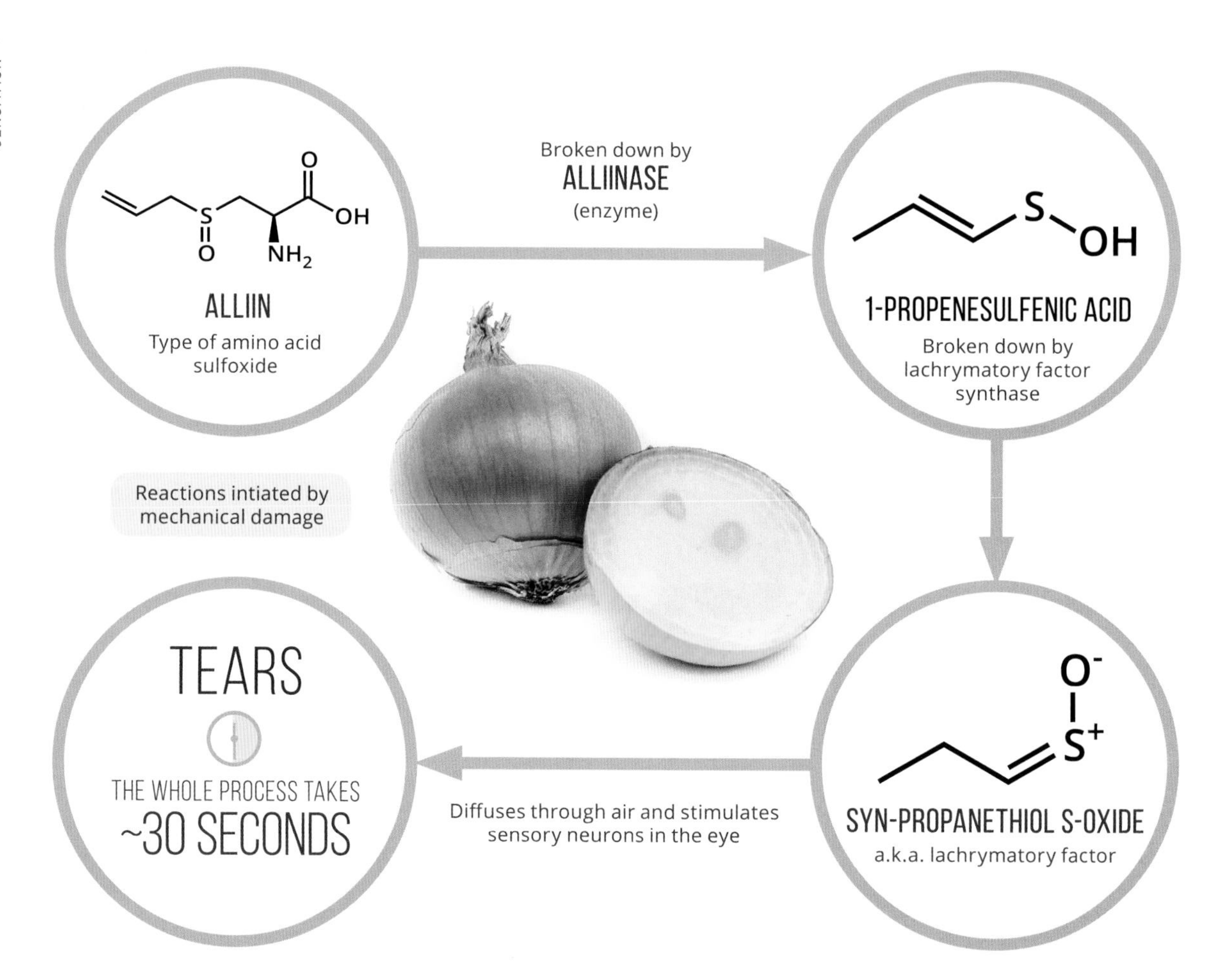
O
OH
S
O
NH_2
ALLIIN
Type of amino acid sulfoxide
Broken down by
ALLIINASE
(enzyme)
S
OH
1-PROPENESULFENIC ACID
Broken down by lachrymatory factor synthase
Reactions intiated by mechanical damage
TEARS
THE WHOLE PROCESS TAKES
~30 SECONDS
Diffuses through air and stimulates sensory neurons in the eye
O^-
S^+
SYN-PROPANETHIOL S-OXIDE
a.k.a. lachrymatory factor

WHY DO ONIONS MAKE YOU CRY?

There can't be many people who haven't experienced the ability of an onion to reduce those chopping it to tears. Interestingly, however, the chemical compounds that cause this effect aren't present at all in an unchopped onion – so where do they come from?

Onions are composed of many different chemicals, including a class of compounds called amino acid sulfoxides. When onions are sliced, the mechanical damage to the onion cells causes the release of a class of enzymes called alliinases from the cells of the onion. These enzymes can break down amino acid sulfoxides into another class of compounds, known as sulfenic acids.

One particular sulfenic acid that can be produced by this process is 1-propenesulfenic acid. This compound can be rapidly converted by another enzyme, called lachrymatory factor synthase, to give the compound syn-propanethial-S-oxide, and it's this compound that gives onions their tear-inducing abilities. Production of this gaseous chemical peaks around 30 seconds after mechanical damage to the onion takes place.

Syn-propanethial-S-oxide causes the production of tears because it can diffuse through the air to your eyes, where it can then stimulate sensory neurons, leading to stinging. The crying response to this is a result of your eye trying to flush the irritant out by producing tears from the tear glands.

There are numerous theories as to how this reaction to chopping onions can be prevented. Personally, I've found the effect is hugely lessened when wearing contact lenses; this is because the lenses sit in front of the cornea, the area which has the highest density of nerve endings in your eye, therefore preventing the chemical coming into contact with them. However, this isn't an option for those who don't wear contact lenses, so what other options are available?

Aside from wearing goggles while chopping onions (fine if you're not averse to looking slightly ridiculous in the comfort of your own home), another suggestion involves putting onions in the fridge, or even the freezer, for 15 minutes prior to chopping. Though this might sound a little odd, scientifically it makes sense: the reactions leading to the production of syn-propanethial-S-oxide will take place more slowly at a lower temperature, and so you can get the onion chopped and into the pan before it can exact its revenge.

WHAT GIVES CHILLIES THEIR SPICINESS?

Chillies come in a huge number of varieties – ghost pepper, cayenne pepper, habanero pepper, jalapeño pepper and serrano pepper, to name but a few. Each of these varieties owes its spiciness to particular compounds, present in varying degrees depending on the strength of the chilli.

A family of compounds called capsaicinoids are responsible for the heat of chillies; within this family, several different compounds are found in different varieties of chillies, but the dominant compound is one called capsaicin. Dihydrocapsaicin is another, similar, compound present in relatively high levels. The ratio of these two compounds varies from chilli to chilli, but together they account for 80–90 per cent of the overall concentration of compounds from the capsaicinoid family.

The capsaicinoids found in chillies bind to a receptor in the mucous membrane of the mouth when ingested; this is the receptor associated with heat and physical abrasion, and hence this produces a burning sensation. Despite this, the compound does not produce any physical or tissue damage. If the compound is ingested repeatedly, the receptors that it binds to can become depleted, effectively allowing you to build up a tolerance. The pain actually produces endorphins, a class of compounds that act as natural painkillers in the body, and can also impart 'a sense of well-being'.

Although no chilli has a capsaicin content high enough to be harmful (even the fabled ghost chillies), capsaicin is nonetheless a toxic compound. As well as its presence in chillies, capsaicin finds a use in pepper sprays in low concentrations as its inflammatory effects cause the eyes to close, incapacitating those it is sprayed at.

The heat of chillies can be measured in a couple of ways. The first method, known as the Scoville scale, is a taste test in which a measured extract of the dried pepper is incrementally diluted with a solution of sugar and water, until the heat is no longer detectable by a panel of five testers. Obviously, this is far from being a precise method (it was devised in 1912).

The other method through which the heat of chillies is measured is the rather more precise procedure of high performance liquid chromatography (HPLC). In this analytical method, a solvent sample is forced through a column under high pressure, to achieve separation of the mixture and determine the capsaicinoid content.

Finally, there's the oft mooted question of how best to soothe the fire of chillies. The long hydrocarbon 'tail' of the capsaicin molecule makes it insoluble in water; it is, however, readily soluble in alcohol and oil. That said, the small percentage of alcohol in beer sadly isn't enough to have much of an impact. The best bet for removing the burning sensation of too much chilli is to drink milk – this contains a class of proteins called casein, which is lipophilic and envelopes the fatty capsaicin molecules, successfully washing them away and preventing them from further stimulating the receptors in the mucus membranes.

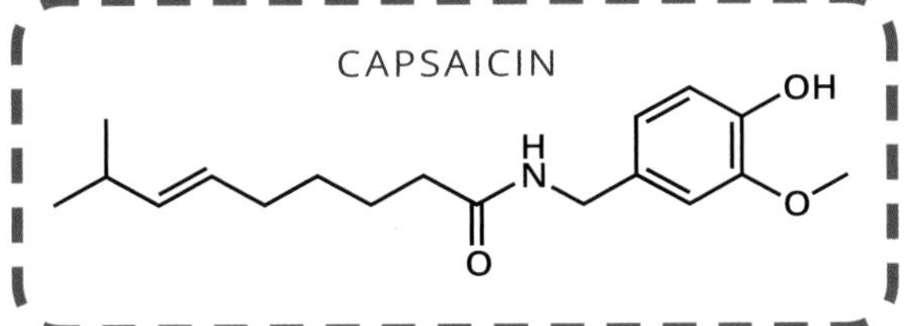

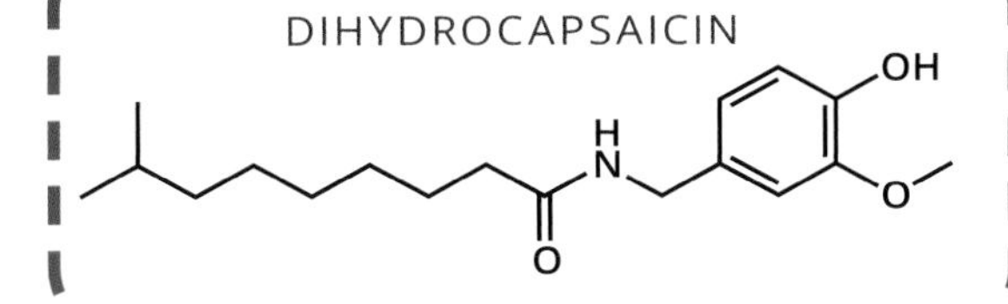

THE SCOVILLE HEAT INDEX

A taste-based scale that measures the pungency of chilli peppers in Scoville Heat Units (SHU).
Increasing concentrations are given to a panel of testers until a majority of the panel can detect the heat.

JALAPEÑO	CAYENNE PEPPER	HABANERO	GHOST PEPPER	PEPPER SPRAY	PURE CAPSAICIN
8,000	50,000	350,000	1,400,000	5,300,000	16,000,000

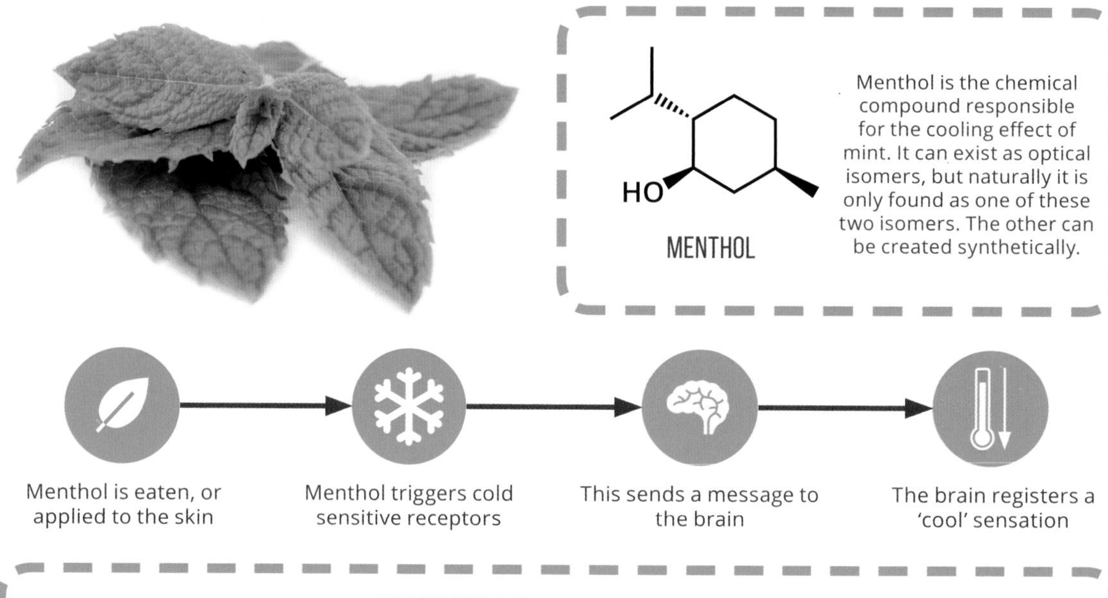

PRODUCTS CONTAINING MENTHOL

A number of commercial products contain menthol, whether for its cooling effect or for its mint flavour.

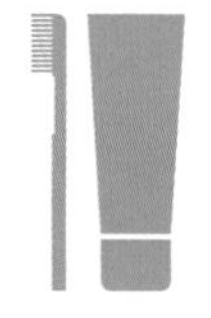
TOOTHPASTES

DECONGESTANTS

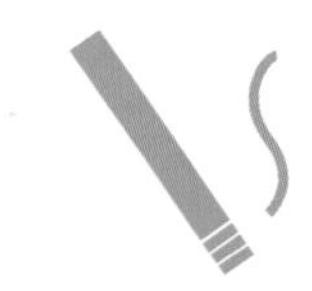
MENTHOL CIGARETTES

TOPICAL ANALGESICS

CHEWING GUM

WHY DOES MINT MAKE YOUR MOUTH FEEL COLD?

A common phenomenon with mint, or mint-flavoured items like chewing gum or toothpaste, is the cooling feeling it seems to generate in your mouth. This isn't all in your mind, but is actually a consequence of one of the chemicals contained in mint – one whose name is likely to be familiar: menthol.

We previously discussed capsaicin, the main 'spicy' compound in chillies, and how it causes the sensation of spiciness by binding to receptors in the lining of the mouth that usually allow us to detect heat. Menthol acts in a similar fashion, by binding to receptors in the mouth responsible for detecting cooler temperatures. It doesn't actually decrease the temperature – it just tricks nerve cells into thinking your mouth is cooler than it is, and this is the message that they relay to your brain.

This cooling effect isn't the only effect that menthol can have on the body, however. Studies have confirmed that it can also have an analgesic, or painkilling effect, and for this reason it is available in a variety of topical creams, gels and even patches for the relief of muscle aches, pains and headaches. The wide range of products it's available in means that you probably use menthol on a daily basis – other products include after-shave creams, decongestant products and mouthwash.

With all of these possible uses, it's perhaps not surprising to learn that the amount of menthol available naturally is far below the demand for it in our everyday lives. The demand is estimated to be as much as 35,000 metric tons of menthol per year. For this reason, the compound is now also prepared synthetically, and has been prepared on an increasingly large scale since 1973.

Since we have previously discussed optical isomerism, it's worth mentioning that menthol also exhibits optical isomerism. It exists naturally only as one optical isomer (it *can* exist as two mirror images of the same molecule, but is only found naturally as one of these mirror images), the form which has a cooling effect. The other isomer, which can be synthesised in a laboratory, interestingly doesn't have anywhere near as pronounced a cooling effect, presumably because the minor difference between the two optical isomers means that it isn't as effective at activating the receptor that detects cooler temperatures.

HOW DOES POPPING CANDY WORK?

You've probably encountered popping candy before – it's available as small, multicoloured clusters, and is also often included in chocolate. As you eat them, they cause loud popping noises to emanate from your mouth. The secret behind how this is accomplished has its roots in the manufacturing process.

Popping candy starts off as a molten mixture of sugars and flavouring. If this mixture is allowed to cool it would produce hard candy, but it wouldn't have the same popping characteristic, so this effect isn't a consequence of the original ingredients. What's still required is a gas we produce in abundance as a result of other processes: carbon dioxide.

The molten sugar mixture is allowed to cool in the presence of high-pressure carbon dioxide; by high pressure, we're talking roughly 50 times the normal atmospheric pressure we experience here on the Earth's surface. This high pressure allows a large amount of the carbon dioxide to dissolve in the sugar, much more than it would be able to at normal atmospheric pressure.

Obviously, when the pressure is eventually removed, the carbon dioxide, still being trapped within the sugar, wants to try to escape. Some of the larger bubbles are able to do so, and crack the solidifying slab of sugar into smaller pieces. However, the smaller bubbles of carbon dioxide get trapped by the solidifying sugar and are unable to escape. This results in as much as 15 centimetres cubed of carbon dioxide remaining trapped per gram of candy. Because of the completely random nature of the fracturing, the exact amount of dissolved gas in each piece will vary significantly. The maximum size of the carbon dioxide bubbles is around 350 nanometres, or 0.00000035 metres.

This is where you come in. When you eat popping candy, the candy dissolves in your saliva, releasing the highly pressurised bubbles of carbon dioxide. This manifests itself in the form of the characteristic popping noises. The carbon dioxide gas released is in no way dangerous – in fact, it's significantly less than the amount of carbon dioxide dissolved in carbonated drinks.

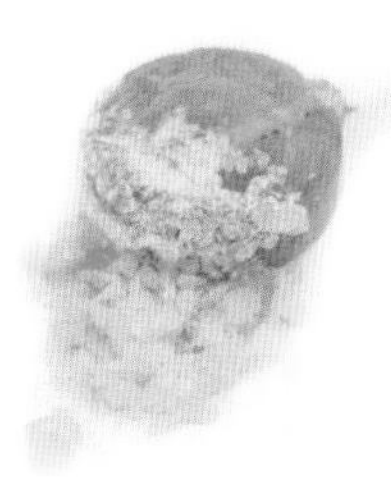

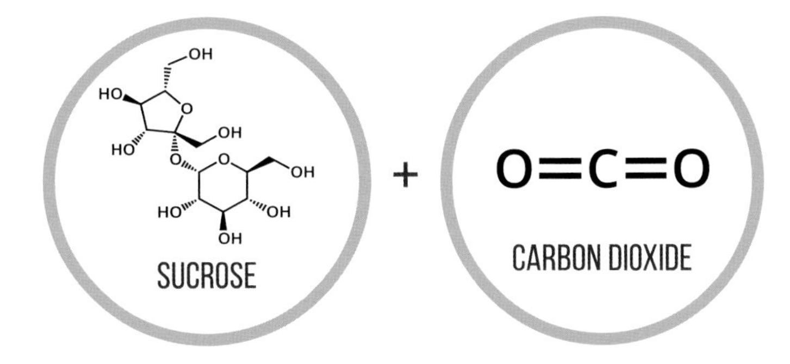

THE MANUFACTURING PROCESS

1

The ingredients of the popping candy are mixed together, then melted at a high temperature, forming a syrup.

2

The syrup is exposed to high-pressure carbon dioxide gas, at roughly 50 times the normal atmospheric pressure.

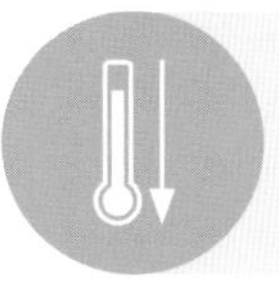

3

The syrup is cooled. Larger bubbles crack the cooling candy, but small, high-pressure-bubbles remain trapped.

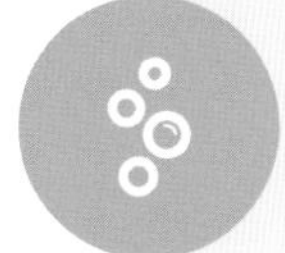

4

Saliva dissolves the candy when it is eaten, releasing carbon dioxide from the bubbles and resulting in popping.

The differing colourations and flavours of popping candy are achieved by the addition of varying chemicals, such as food colourings.

WHAT CAUSES THE PUNGENCY OF WASABI?

Anyone who's a fan of sushi will be familiar with the pungent kick of wasabi. Usually served up in the form of a light green paste, it's actually derived from the root of the wasabi plant, also referred to as Japanese horseradish. Its strong flavour is often described as hot – but this is a different heat to the spiciness of the chilli, affecting the nasal passages more than the tongue.

Wasabi is prepared from the grated root of the plant, and the compounds that cause the pungency are produced when the tissue of the plant is mechanically damaged. Compounds called glucosinolates are broken down by enzymes, and then react further to produce a number of compounds called isothiocyanates. The major compound responsible for wasabi's pungency is allyl isothiocyanate, which constitutes around 100 milligrams per 100 grams.

Other compounds, however, also contribute to the 'fresh green' taste of wasabi, though they are present in much smaller amounts. 6-methylthiohexyl isothiocyanate is thought to contribute a freshness to the flavour, while 7-methylthio-heptyl isothiocyanate contributes a sweetness to the taste, and 8-methylthiooctyl isothiocyanate contributes a weaker pungency. While normal horseradish also has a certain pungency to its flavour, and allyl thiocyanate is found in both wasabi and horseradish, it's interesting to note that the methylthioalkyl thio-cyanates are only found in appreciable concentrations in wasabi.

The isothiocyanates generated by the enzymatic breakdown of glucosinolates are very volatile, meaning they can escape as a gas very easily; this volatility is probably a key reason why the sensation of eating wasabi is felt strongly in the nasal passages. The chemicals are detected by receptors on some nerve cells, known as TRPA1, which transmits sensation to the brain, leading to wasabi's curious effect.

The volatility of the isothiocyanates is also the reason that wasabi usually has to be freshly prepared, as left exposed to the air the compounds slowly evaporate from the paste. Wasabi can be freeze-dried or air-dried in order to try to combat this problem, and can be reconstituted with water when required.

The properties of isothiocyanates may also lend themselves toward medical uses. There's some research that suggests they may have applications as mild anti-inflammatory com-pounds – unsurprisingly, horseradish is a traditional remedy for congested sinuses, and it seems this may have a scientific basis. Research on animals also suggests that the some isothiocyanates could have a preventative effect against breast, stomach and colon cancer, although the particular isothiocyanates identified in the research are not those commonly found in wasabi.

GLUCOSINOLATES

broken down to produce isothiocyanates

ISOTHIOCYANATES

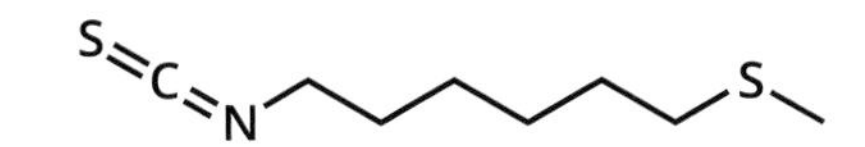

6-METHYLTHIOHEXYL ISOTHIOCYANATE

freshness

7-METHYLTHIOHEPTYL ISOTHIOCYANATE

sweetness

8-METHYLTHIOOCTYL ISOTHIOCYANATE

a weak pungency

ALLYL ISOTHIOCYANATE

major isothiocyanate compound; largely responsible for wasabi's pungency

MIND

TURKEY & TRYPTOPHAN

Tryptophan is an essential amino acid for humans – that is, it cannot be synthesised by the human body, and must instead be obtained from food.

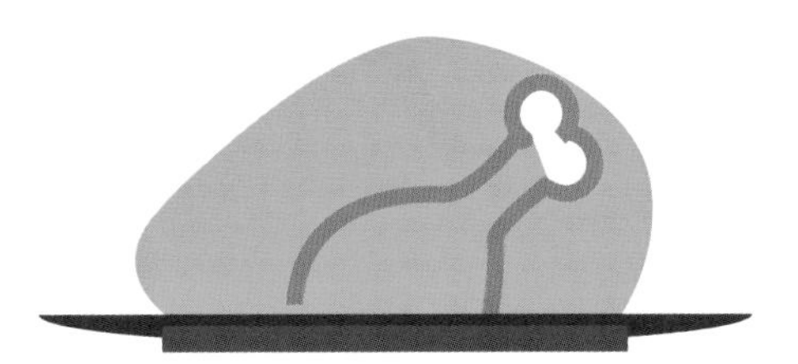

H N NH_2 OH O

TRYPTOPHAN

H N HO NH_2

SEROTONIN

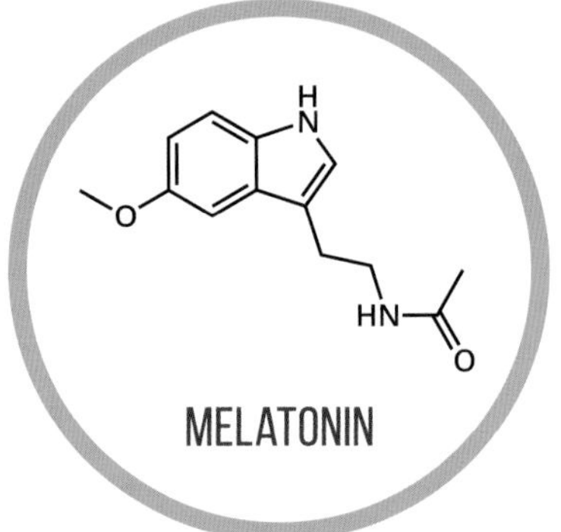

OTHER FOODS THAT CONTAIN TRYPTOPHAN

CHEDDAR CHEESE
0.32g per 100g

SALMON
0.22g per 100g

EGG
0.17g per 100g

MILK
0.08g per 100g

DOES EATING TURKEY MAKE YOU SLEEPY?

At Christmas in the UK, or Thanksgiving in the USA, the centrepiece of the day's meal will commonly be a large, family-sized turkey. You can perhaps stave it off in your earlier years, but as you get older the prospect of a post-dinner nap becomes all the more appealing. It's commonly stated that the catalyst of this sudden sleepiness is the chemical tryptophan, found in turkey meat (around 0.3g per 100g), but is there any truth to this?

Tryptophan is an amino acid; amino acids are the building blocks for the creation of proteins in the body. One of a number of amino acids that we don't produce naturally in our bodies, tryptophan must therefore be obtained from our diet. It's found in high levels in meat, fish, dairy products, nuts and seeds. In the body, tryptophan can be utilised to synthesise serotonin. Serotonin is a neurotransmitter, a chemical that performs a number of roles in the brain. It affects mood, contributing to feelings of well-being and happiness, and low levels have been linked to depression. It is also needed to make the chemical melatonin, which is important in regulating our sleep cycles.

With serotonin being something of a calming agent, you might think it's a logical conclusion that taking in more tryptophan could make you more sleepy. Plenty of people were convinced of this – in the 1980s, some people were even taking tryptophan supplements to combat insomnia. Unfortunately, it's not quite that simple.

In order to be used to make serotonin, tryptophan has to cross from the bloodstream into the brain. The brain doesn't let just any old chemicals in, and the blood–brain barrier exists to allow the passage of required substances in and out, while preventing neurotoxic substances passing in. Any amino acids that want to pass into the brain have to hitch a ride on transport proteins whose job it is to facilitate this transfer. Turkey contains lots of amino acids, as well as tryptophan, and they all compete for a place on these transporter proteins. Additionally, tryptophan is actually at a much lower concentration than the other amino acids, so it tends to lose out, and as such not a great deal of it will succeed in crossing the blood–brain barrier.

It's safe to conclude then, that turkey isn't the cause of any drowsiness. More likely is the simple fact that you've eaten a lot more food – studies have shown that eating large amounts of carbohydrates can increase serotonin levels in the brain, irrespective of tryptophan content. Also, if it's a celebratory meal, you've probably knocked back an alcoholic drink or two, and these are also much more likely causes of your drowsiness than tryptophan in turkey!

CAN CHEESE REALLY GIVE YOU BAD DREAMS?

A common cautionary tale with regard to cheese is that it can give you nightmares, or at the very least vivid dreams, if you eat it soon before going to sleep. This perception is so widespread that the British Cheese Board, a cheese industry group who proclaim themselves 'the voice of British cheese', even carried out a study to try to disprove it. So, what did they find?

In their study, which involved 200 volunteers, they got subjects to eat a small amount of cheese about half an hour before bed, and then record their dreams. You can certainly question the validity of the study – they tested different types of cheese, but there doesn't seem to be any mention of a control group who weren't eating cheese to compare against. Nonetheless, the results make for some entertaining reading.

They found that blue cheese produced particularly vivid dreams; one subject related a tale of a vegetarian crocodile who was upset that he couldn't eat children, while another dreamt of soldiers engaging in warfare, armed not with guns, but with kittens. Cheddar apparently made people dream of celebrities, red Leicester produced nostalgic dreams, Lancashire cheese invoked dreams of work and Cheshire cheese produced no dreams at all.

It's pretty clear that this research shouldn't be taken too seriously – after all, the British Cheese Board probably have a minor vested interest in people not thinking cheese causes nightmares. There is no scientific data to back up the claim that cheese can cause nightmares, or even alter the content of your dreams. There are, however, some theories about its compounds that could cause unusual effects during your slumbering hours.

Cheese contains high levels of tyrosine, an amino acid. If you're a fan of aged cheeses, such as Parmesan, Gouda or Gruyère, you might have noticed small, white, crunchy crystals in the cheese. These crystals are actually clusters of tyrosine, formed as protein chains in the cheese that unravel with ageing. In the body, tyrosine is converted to tyramine, a chemical which can stimulate the release of the neurotransmitters norepinephrine and epinephrine (adrenaline). The release of these could potentially cause disturbed sleep, which in turn is likely to lead to vivid dreams. That said, it's unlikely that eating a small amount of cheese shortly before sleep will provide enough tyrosine in order to induce these effects.

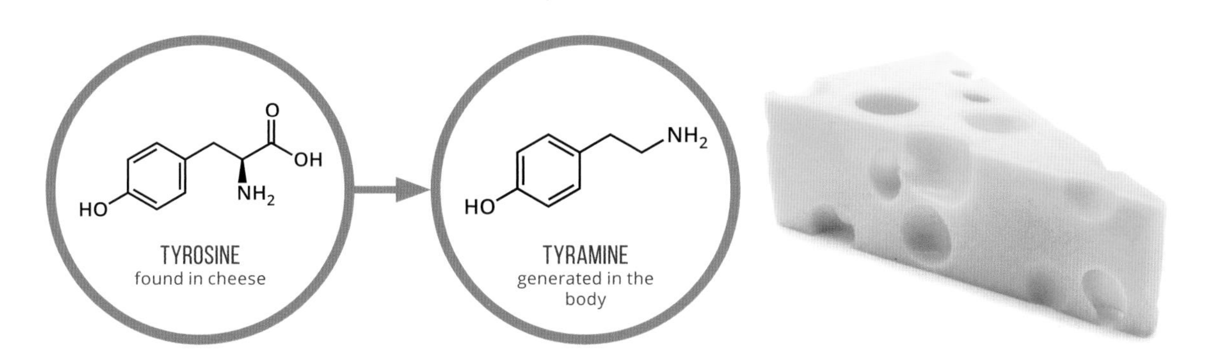

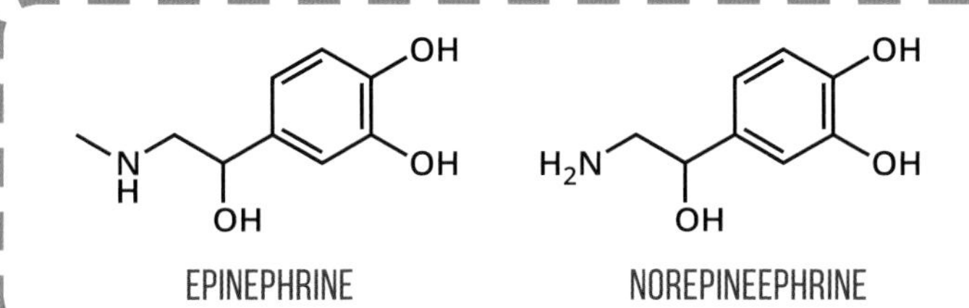

THE EFFECTS OF TYRAMINE

It's suggested tyramine can lead to higher levels of the neurotransmitters epinephrine and norepinephrine, which could cause the effect of bad dreams. However, it's unlikely that the levels of tyramine in cheese are high enough.

APPROXIMATE TYROSINE CONTENT OF DIFFERENT CHEESES

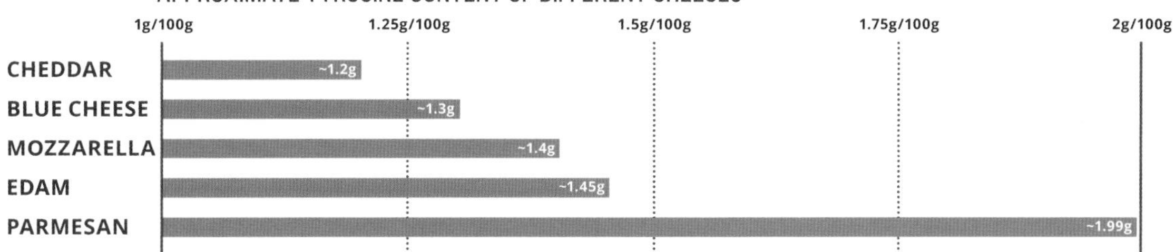

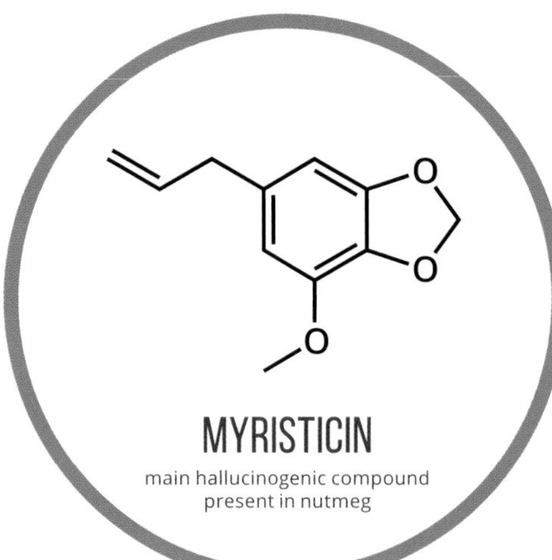

MYRISTICIN

main hallucinogenic compound present in nutmeg

1 TABLESPOON

is enough to induce unpleasant symptoms

SYMPTOMS

NAUSEA

HALLUCINATIONS

RAISED HEART RATE

Other reported effects of nutmeg ingestion include vomiting, euphoria, flushing and a dry mouth.

DURATION EFFECTS PERSIST FOR 24–36 HOURS AFTER INGESTION

Anecdotal evidence suggest that after-effects of excessive nutmeg ingestion can, in some cases, be felt up to a week after ingestion.

Other compounds in nutmeg are also thought to contribute partially to the hallucinogenic effect

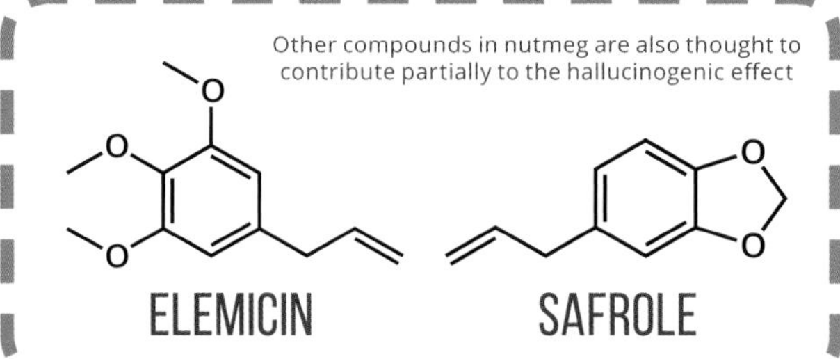

WHY CAN NUTMEG ACT AS A HALLUCINOGEN?

When you think of hallucinogens, you probably wouldn't expect to find one lurking, unbeknownst to you, in your kitchen spice rack. However, the hallucinogenic properties of nutmeg have been known for some time – historical records as far back as the sixteenth and seventeenth centuries comment on its narcotic effects. So, what are the chemical compounds that cause this?

Several compounds have been implicated in the hallucinogenic effect of nutmeg, the main one being myristicin, which accounts for approximately 1.3 per cent of raw nutmeg. It has been suggested in research that the effects of nutmeg could be due to the breakdown of myristicin in the liver into MMDA, a drug of the amphetamine class and known psychedelic. However, while this transformation has been observed in the livers of rats, there has been no evidence of such a transformation occurring in humans.

Interestingly, when a significant amount of pure myristicin was given to a group of subjects (twice the amount present in 20 grams of nutmeg), while six out of ten showed some effects, they were much milder than expected in comparison to the effects of nutmeg. This suggests that the presence of other compounds in nutmeg must also be important in inducing the full 'nutmeg effect'. Compounds that are suspected of contributing to the effect are elemicin and safrole.

Before you reach for an experimental spoonful of nutmeg, it's worth noting the effects it can induce. 1–2 milligrams of nutmeg per kilogram of body weight can induce effects in the central nervous system (myristicin inhibits nerve impulses responsible for involuntary movement of muscles in certain systems in the body, such as the gastrointestinal tract and lungs), and anecdotal records state a tablespoon is enough to bring on other effects including nausea, vomiting, flushing, elevated heart rate, euphoria, hallucinations and a dry mouth; on the face of it, not a particularly cheery band of side effects.

It doesn't really get much better – as well as some of the effects being less than pleasant, they can last for several days, with some reporting symptoms such as vision, balance and concentration problems lasting for over a week. In all, it's probably best that the nutmeg stays confined to your kitchen spice rack.

WHY DO THE STIMULANT EFFECTS OF TEA AND COFFEE DIFFER?

Tea, like coffee, contains caffeine. Obviously, the caffeine in tea has the exact same effect on the brain as the caffeine in coffee, being the same chemical molecule and all. However, most people would agree that the stimulant effect of coffee is a lot more potent than that of tea. Is it simply a lower level of caffeine, or is there something else going on chemically?

Firstly, caffeine levels in tea are undoubtedly lower than in coffee. Obviously these levels will vary from coffee to coffee and tea to tea, but in general a cup of tea contains around half the caffeine of a same-sized cup of coffee. Caffeine works by blocking receptors that the brain chemical adenosine usually binds to induce tiredness. Logically, then, you'd expect the stimulant effect of tea to be weaker than that of coffee – less caffeine will be blocking access to the receptors to which adenosine binds.

However, the key chemical when it comes to the difference between tea and coffee may not be caffeine, but L-theanine, an amino acid commonly found in tea but not coffee. A cup of black tea, on average, contains around 25 milligrams of L-theanine. Measurements of electrical activity within the brain after subjects took L-theanine supplements suggested that it's capable of relaxing the mind, but without causing any drowsiness. This study was carried out at higher L-theanine levels than found in tea, but further studies found the effect was still present at levels akin to normal dietary intake.

Further studies went on to look at the effect of L-theanine in combination with the amount of caffeine. They found that subjects in their study completed tasks quicker and more accurately after ingesting a drink containing a combination of L-theanine and caffeine, compared to a drink containing caffeine, or a placebo. They were also judged to be less susceptible to distracting information provided during memory tests. One caveat is that this study used levels of caffeine and L-theanine slightly higher than those in a single cup of tea, which obviously could have some bearing on their results.

That said, it seems clear that L-theanine does have effects on the mind, and that these are still likely to be present to some extent at the levels of L-theanine found in tea. Further studies will likely tell us more about the difference in the stimulant effects of tea and coffee, but in the meantime, if you're looking for a smoother caffeine hit, tea seems to be the way to go.

COFFEE

Average values for a 200ml cup of coffee

TEA

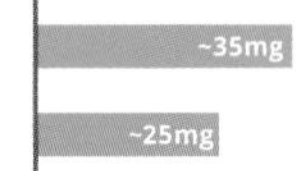

CAFFEINE

L-THEANINE

Average values for a 200ml cup of black tea

O N N O N N

CAFFEINE

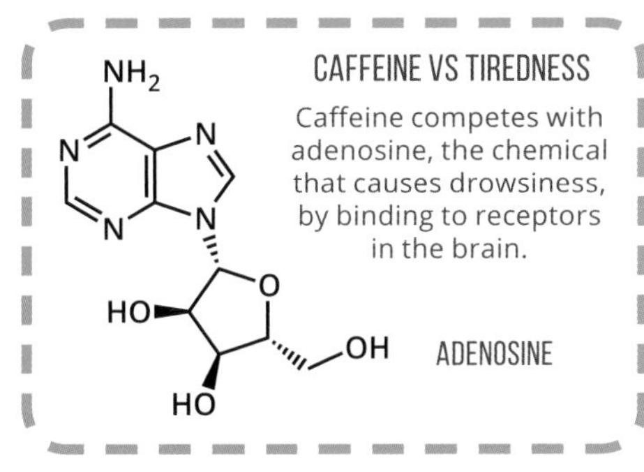

CAFFEINE VS TIREDNESS

Caffeine competes with adenosine, the chemical that causes drowsiness, by binding to receptors in the brain.

O O N H OH NH_2

L-THEANINE

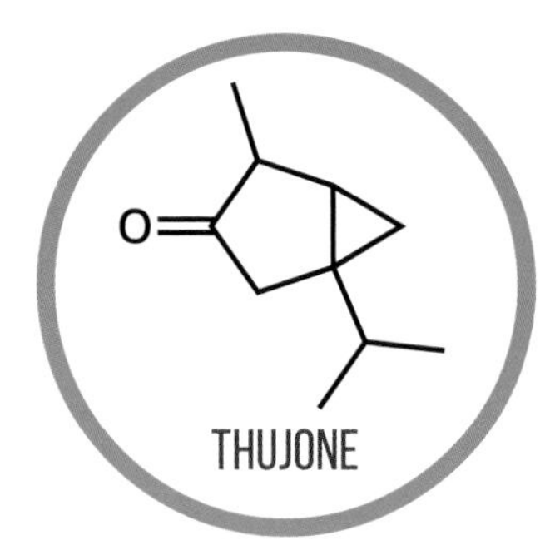

THUJONE – HALLUCINOGEN?

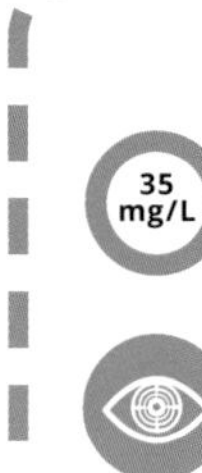

Legal limit on thujone concentration in absinthe. It's thought that even pre-ban absinthe largely conformed to this.

There is no conclusive evidence for psychotropic action of thujone. In amounts found in absinthe, it will not induce hallucination.

A TIMELINE OF ABSINTHE CONTROVERSY

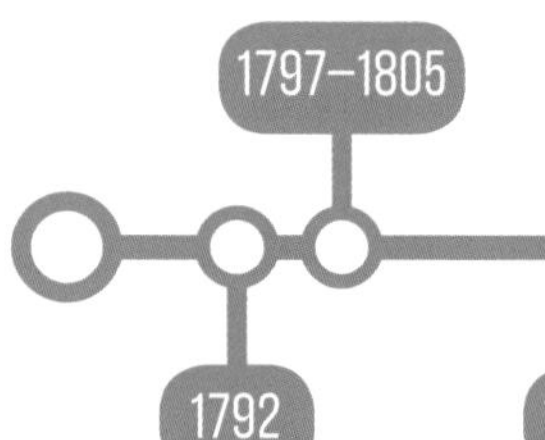

1792 Dr Pierre Ordinaire creates the first recipe for modern distilled absinthe.

1797–1805 Henri-Louis Pernod opens a number of absinthe distilleries in Switzerland and France.

1840s Absinthe given to French soldiers as a malaria preventive.

1860s Absinthe's popularity soars in France; the hour from 5 p.m. is known in bars as 'the green hour'.

1864–74 Magnan's studies on 'absinthism'.

1905 Swiss farmer murders family after drinking absinthe, leading to petition to ban.

1906–14 Banned in Belgium, Brazil, the Netherlands, Switzerland, the United States and France.

1990s Absinthe revival in countries such as the UK, where it was never banned.

2007 France becomes the last major country to repeal its ban on absinthe.

DOES DRINKING ABSINTHE CAUSE HALLUCINATIONS?

Absinthe is an anise-flavoured spirit made from a blend of herbs and wormwood. As well as being obviously alcoholic, with an alcohol percentage that can be as high as 90 per cent in some cases, absinthe has a persisting reputation as a hallucinogenic drink. For much of the twentieth century, it was banned in the USA and a large part of Europe, though in the present day these bans have been repealed. The reason for the bans was absinthe's suspected hallucinogenic properties.

These properties were suggested as early as the nineteenth century, by a French physician, Valentin Magnan. He suggested absinthe causes a syndrome, referred to as 'absinthism', which caused addiction, seizures and hallucinations. He singled out wormwood as the culprit, and used evidence of wormwood's convulsive effect on test animals to back his case.

Wormwood, a key ingredient in absinthe, contains a particular chemical, thujone, which has known convulsant effects. Magnan eventually isolated thujone and identified it as the cause of 'absinthism'. However, in the tests on guinea pigs that he used as evidence, the animals were given wormwood extract which contained levels of thujone many times higher than those found in absinthe. This was enthusiastically pointed out by Magnan's critics at the time, but nonetheless his work laid the foundations for the ban on absinthe in many countries that remained in place for decades.

Levels of thujone in absinthe today are subject to regulation – for example, in the EU it is limited to only 10 milligrams per litre. Some manufacturers still try to play on the supposed hallucinogenic properties of the drink in their advertising, but there is absolutely no evidence that the ingestion of thujone leads to hallucinations. Even in doses much higher than those found in absinthe, this would not be expected. The dose at which convulsions would be seen is also impossible to reach by drinking absinthe without first succumbing to alcohol poisoning.

We know, then, that thujone levels in absinthe today aren't a problem. But surely it's possible that the levels before this regulation were much higher, and could have caused the absinthism that Magnan observed? In order to prove that this was not the case, in a research study scientists obtained 13 bottles of absinthe from before the 1915 ban in France, and tested them for their thujone content. They found that, at highest, the concentration of thujone was 48.3 milligrams per litre, still far too low to cause convulsions. They also ruled out the possibility that the thujone content of the bottles had changed with age. They found nothing that could cause the observed symptoms of absinthism other than the ethanol present.

In short, then, it seems that the symptoms of 'absinthism' observed by Magnan in human subjects were little more than misidentified alcoholism. There is no evidence that thujone has any hallucinogenic properties, and thus there is no possibility that absinthe can provide any drug-like highs.

HOW DO ENERGY DRINKS WORK?

Over the past decade or so, the energy drinks market has been booming, with a large number of different brands now available in shops. They're marketed as providing a boost to both physical and mental function – but how much of this is true, and how much is just propaganda?

For starters, the term 'energy drink' is really something of a misnomer. While the sugar content of these drinks certainly does provide energy, in most cases the sugar levels are comparable to those in other soft drinks which are not marketed as energy drinks. The other active chemical ingredients present can have some effects, but are not a source of energy in the strictest sense of the word.

Apart from sugar, the other main constituent of energy drinks which contributes towards their effect is caffeine. We already discussed the effects of caffeine and how it binds to adenosine receptors in the brain, preventing the slowing of neural activity. Many energy drinks tend to contain 80 milligrams of caffeine per serving; the amount of caffeine in these drinks isn't specifically limited by law, however, so some may contain more. The FDA has stated that a daily caffeine intake of 400 milligrams has not been associated with negative health impacts, although other studies suggest that temporary adverse effects, such as headaches and insomnia, can sometimes be seen with an intake greater than 200 milligrams.

Caffeine, then, has proven effects on the brain, and likely contributes improved alertness after consumption of an energy drink. However, many energy drinks include other ingredients as well, and one of the most popular inclusions is taurine. Taurine is a compound originally isolated from ox bile, and is also a major constituent of human bile, but these days the taurine in energy drinks is produced entirely synthetically in a lab. In our body it has several important biological roles (including cardiovascular function, the central nervous system and skeletal muscles), but its efficacy in energy drinks is subject to debate.

Much research on taurine as a supplement seems to be focused on its presence in energy drinks, and these are often used as the test substances in studies on its effects. However, as the energy drinks also contain caffeine, it is difficult to ascribe any effects to taurine independently if they are both taken together. Most studies have concluded that the cognitive effects of energy drinks are owed mainly to caffeine, and there is little scientific evidence that the wide range of other ingredients have any discernible effect. With that in mind, a cup of coffee, containing a comparable amount of caffeine, would seem to be just as effective in increasing your alertness as a can of an energy drink.

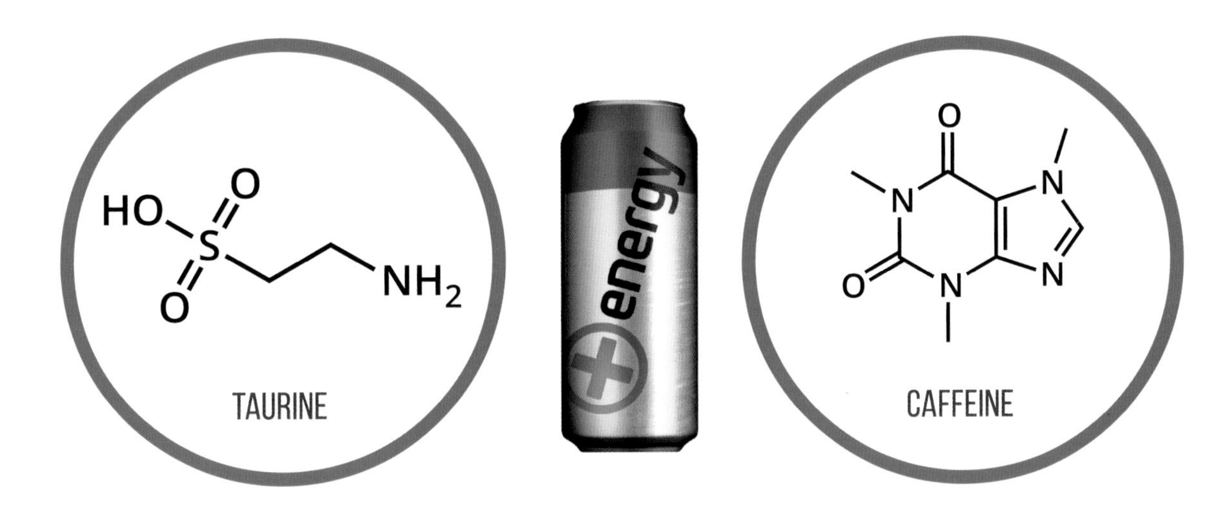

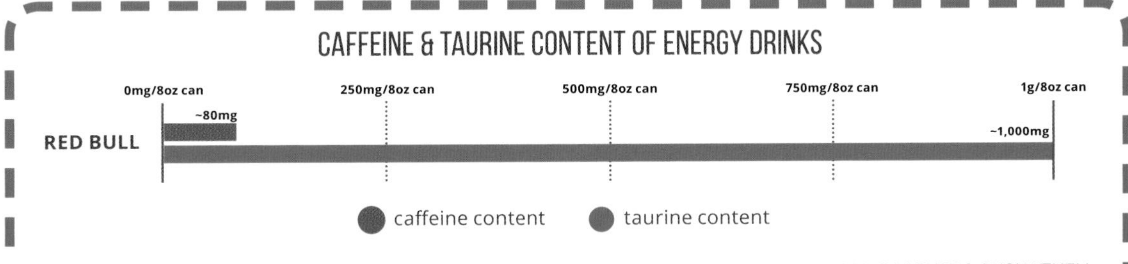

SOME STUDIES SHOWED ENERGY DRINKS HAVE MODEST EFFECTS ON ENDURANCE, BUT MANY OTHERS SHOW THEY HAVE NO EFFECT. ANY EFFECT THAT THEY DO HAVE IS LIKELY DUE TO CAFFEINE CONTENT RATHER THAN TAURINE.

HEALTH

85+ DRUGS

with known grapefruit interactions

ACEBUTOLOL ALISKIREN
AMITRIPTYLINE AMIODARONE
AMLODIPINE AMPRENAVIR
APIXABAN ATORVASTATIN
BUDESONIDE BUSPIRONE
CAFFEINE CARBAMAZEPINE
CARVEDILOL CILOSTAZOL
CISAPRIDE CLARITHROMYCIN
CLOMIPRAMINE CLOPIDOGREL
COLCHICINE CRIZOTINIB
CYCLOSPORINE DARIFENACIN
DASATINIB DEXTROMETHORPHAN
DIAZEPAM DIGOXIN
DILTIAZEM DOMPERIDONE
DRONEDARONE EPLERENONE
ERLOTINIB ERYTHROMYCIN
ESTROGENS ETOPOSIDE
EVEROLIMUS FELODIPINE
FENTANYL FESOTERODINE
FEXOFENADINE FLUVOXAMINE
ITRACONAZOLE LAPATINIB

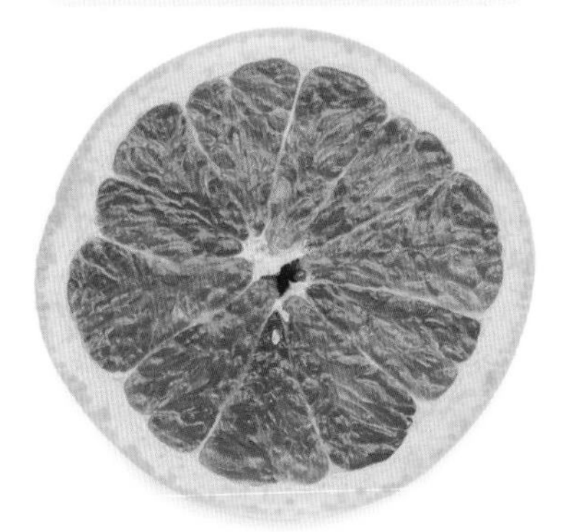

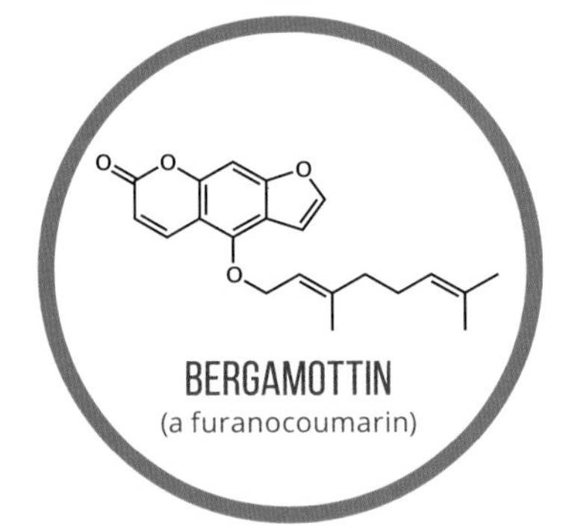

BERGAMOTTIN
(a furanocoumarin)

LEVOTHYROXINE LOSARTAN
LOVASTATIN LURASIDONE
MARAVIROC METHADONE
METHYLPREDNISOLONE
MIDAZOLAM NICARDIPINE
NIFEDIPINE NILOTINIB
NIMODIPINE NISOLDIPINE
OXYCODONE PAZOPANIB
PIMOZIDE PRIMAQUINE
PROGESTERONE QUAZEPAM
QUETIAPINE QUININE
RILPIVIRINE RIVAROABAN
SAQUINAVIR SCOPOLAMINE
SERTRALINE SILDENAFIL
SILODOSIN SIMVASTATIN
SIROLIMUS SOLIFENACIN
SUNITINIB TACROLIMUS
TAMSULOSIN THEOPHYLLINE
TICAGRELOR TRIAZOLAM
VANDETANIB VERAPAMIL
WARFARIN ZIPRASIDONE

Drugs underlined are those with which a major interaction is seen, and with which ingesting grapefruit is not recommended. This is not intended to be a comprehensive list, and may not include all drugs for which interaction is seen with grapefruit.

WHY CAN'T YOU EAT GRAPEFRUIT WITH SOME MEDICATIONS?

You may have heard of 'the grapefruit juice effect'. A range of medications recommend avoiding grapefruit (and grapefruit juice) while you are taking them, due to the side effects that this can cause. The side effects are due to the presence of particular chemicals within the grapefruit.

The principal culprits are a family of chemical compounds called furanocoumarins, and particularly the compounds bergamottin and dihydroxybergamottin. Both of these compounds interfere with the activity of an enzyme that plays an important part in breaking down some drugs in the body, preventing it from doing so. This, in turn, can lead to increased levels of concentration of the drug in the bloodstream. This is a problem, because prescriptions for drugs take into account the rate at which your body breaks down the drug in their dosage recommendations. Since these compounds in grapefruits usually greatly decrease the rate of breakdown, repeated doses can lead to much higher doses of the drug in the bloodstream, in turn potentially leading to harmful side effects.

This effect of grapefruit juice is also long lasting, with it taking around 24 hours for the enzyme's activity to recover to half of its original level, and full recovery taking up to 72 hours. One whole grapefruit, or 200 millimetres of juice, can be enough to cause significant interaction with enzyme activity. The side effects can vary depending on the drug being taken, but can potentially include kidney damage, blood clots and breakdown of muscle fibres.

The pomelo, a fruit which is a hybrid of an orange and a grapefruit, exhibits similar interactions with the enzyme; however, the tangelo, a fruit which is a cross between a tangerine or pomelo and a grapefruit, contains only trace amounts of bergamottin, and as such will not interact with the enzyme and can be consumed safely with the same drugs that grapefruit would interfere with.

As an interesting aside, grapefruit's negative side effects on drugs have also been investigated for their potential benefits. Several anti-AIDS drugs which would otherwise be broken down in the body quite quickly, could have their usefulness extended by being taken in combination with grapefruit, which would extend the amount of time the drug is present in the bloodstream.

WHY DO LEMONS HELP PREVENT SCURVY?

Lemons contain a number of acids; the major acidic compound, citric acid, likely needs no introduction, and even has its own E number (E330). There are, though, a couple of other acids found in a lemon's chemical structure that make important contributions. One of them is the reason why lemons are recommended as a way of warding off a specific disease, scurvy.

Citric acid is the main contributor to the lemon's sour taste. Malic acid is present in around 5 per cent of the concentration of citric acid. It, too, has its own E number (E296), and is also found in apples and cherries, where it is responsible for aspects of their flavour.

Another acid present in lemons, and one with which citric acid is occasionally confused, is ascorbic acid, or vitamin C. The vitamin C levels in a lemon, at around 50 milligrams per 100 grams, are on a par with those of an orange, and significantly higher than those in a lime (~29mg/100g). This last fact in particular is one that the British Navy discovered belatedly to their detriment in the early 1900s.

Vitamin C is required by the body to produce collagen, the main protein of connective tissues in animals. Scurvy is a disease caused by a lack of vitamin C, the symptoms of which include spots, bleeding gums, loss of teeth, jaundice, fatigue, joint pain, fever and eventual death. The disease was a major problem for seafarers, who would spend months at sea, and without a supply of fresh citrus fruit to supplement their vitamin C levels often succumbed to scurvy. By the mid-1700s, physicians had, however, discovered that citrus fruits were an effective cure for the disease, and in the late 1700s all Royal Navy ships were required to serve lemon juice in rations.

Despite this recommendation, a lack of awareness of vitamin C, and the differing vitamin C content of lemons and limes, meant scurvy again became an issue in the early 1900s. The Royal Navy began to start using lime juice instead of lemon, as they could source these from within the British colonies. They did so under the assumption that the acidity of lemons was what warded off scurvy, and as limes were more acidic it followed that they would be equally effective. This had occasionally dire consequences, with several Arctic expeditions succumbing to scurvy due to the failure of lime juice to provide enough vitamin C.

The confusion this caused was not fully resolved until the eventual isolation and discovery of vitamin C by the Hungarian Albert Szent-Györgyi in 1932. Vitamin C was actually named after its scurvy-preventing abilities – the name, 'ascorbic acid', comes from 'antiscorbutic', a term used to refer to substances preventing scurvy.

SCURVY THE SYMPTOMS

After 3 months of vitamin C shortage

FATIGUE

JOINT PAIN

RED-BLUE SPOTS

BLEEDING GUMS

SHORTNESS OF BREATH

JAUNDICE

A TIMELINE OF SCURVY BETWEEN 1500–1800, IT'S ESTIMATED SCURVY KILLED 2 MILLION SAILORS

1500 (BC)
First known written record of scurvy

1499
Vasco da Gama loses 116 of crew of 170, many to scurvy

1520
Ferdinand Magellan loses 208 of a crew of 230, mainly to scurvy

1747–62
James Lind trials show lemon juice prevents scurvy

1795
British Navy makes use of lemon juice mandatory

1932
Vitamin C linked with scurvy prevention

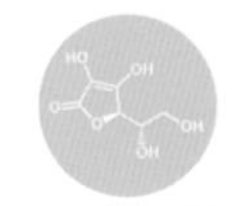

THE ALLERGIC RESPONSE

1. Nut protein exposure results in the body misidentifying them as a threat; antibodies released to combat them.

2. The antibodies produced bind to two types of cell in tissues – mast cells and basophils.

3. These cells release a number of chemicals, including histamine, which produce an inflammatory response.

4. In severe cases, symptoms of this response include vasodilation that can lead to anaphylactic shock.

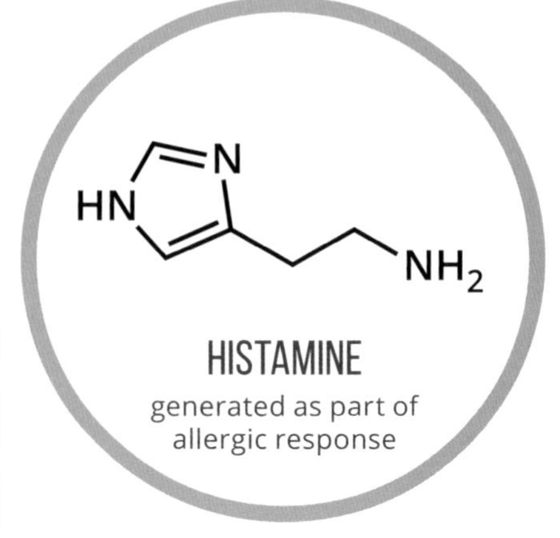

HISTAMINE

generated as part of allergic response

1.3% estimated US adults with a nut allergy

TREE NUT ALLERGY

almonds, Brazil nuts, cashews, chestnuts, hazelnuts, macadamia nuts, pecans, pistachios, walnuts

PEANUT ALLERGY

distinct from tree nuts, as peanuts are legumes; however, many are allergic to both tree nuts and peanuts

WHY ARE SOME PEOPLE ALLERGIC TO NUTS?

These days, foods in supermarkets must be clearly labelled as to whether or not they contain nuts. If there's even the slightest possibility that they do, they must bear that fact clearly on their packaging. But what are the chemical processes that lead to the allergic reaction to nuts in foods?

It's estimated that as much as 1–2 per cent of the general population has a nut allergy, and this figure is thought to be on the rise. The allergic reactions to nuts, or products containing them, can be severe and life-threatening. The exact cause of someone developing a nut allergy is still unknown; it can be hereditary, but it's not always passed on. What we do know is that specific proteins found in nuts can trigger allergic reactions due to the body's response.

The exact protein will vary from nut to nut, so it's not a case of it being one universal protein that triggers allergies. The allergic response is a result of the body mistakenly identifying a benign substance as a threat, and the response itself is the same no matter what the allergen. When you're first exposed to an allergen, your body produces antibodies specifically designed for it which bind to cells called mast cells, as well as others called basophils. At this stage, you still won't experience any effects; however, when exposed to more of the allergen, the antibodies bound to the mast cells now bind to molecules of the allergen; when enough antibodies are bound between the two, the mast cells explode, releasing a number of chemicals which includes histamine.

Histamine produces an inflammatory response – this can cause swelling, sneezing and itching. It's also the chemical that's largely responsible for the symptoms of hay fever. In severe cases of nut allergy it can lead to anaphylactic shock, which results from the dilation of blood vessels and the subsequent lowering of blood pressure. The main treatment for anaphylactic shock is administration of epinephrine, more commonly known as adrenalin, which causes blood vessels to constrict, reversing the dilating effect of the allergic response.

You might think that those with nut allergies simply have to avoid ingesting or coming into contact with nuts themselves. However, a recent news story suggests even more care may be necessary if you're allergic to Brazil nuts. Research documents the case of a UK woman in 2006, who developed an allergic reaction after intercourse with her boyfriend. She had not ingested nuts prior to this, but her boyfriend had. It became apparent that the protein responsible for triggering the allergic response must have been able to pass through his digestive system without breaking down and pass into his semen. Allergy tests with his semen confirmed this, making it the first documented case of allergy transmission via intercourse. Unfortunately, the couple broke up before any further research could be carried out.

CAN A TICK BITE MAKE YOU ALLERGIC TO MEAT?

Allergies are a part of many people's everyday lives, be it nut allergies or hay fever. The stimulus differs between allergies, but the response of the body is the same as that we discussed when looking at nut allergies. An odder allergy, which it is estimated affects thousands of people, is an allergy to red meat.

This allergy was first described back in 2007, and to understand its origins we need to look to a particular species of tick. The lone star tick is found in approximately 28 states in the USA, predominantly on the eastern side of the country, and has been linked with the development of red meat allergy. The trigger for the allergy is a particular chemical compound: a sugar known as alpha-gal.

Alpha-gal is produced by most mammals and is commonly found in their cell membranes. Humans and primates, however, are an exception to this – they don't produce alpha-gal in their cells. Normally, alpha-gal in our digestive system isn't a problem, but if it's introduced to our bloodstream it's a different story. When someone is bitten by a lone star tick, alpha-gal can be transferred to the unfortunate victim. Because alpha-gal isn't a compound that's produced by our bodies, it's treated as an interloper by our immune systems, and antibodies are produced in order to combat it. It's these antibodies that can then cause the allergic reaction to red meat.

Red meat also contains alpha-gal, and its presence can trigger the antibodies that were produced in response to the alpha-gal transferred by the tick bite. The reaction isn't instantaneous – it usually occurs between three to seven hours after ingestion of red meat, and can lead to hives, inflammation, vomiting, diarrhoea and potential anaphylactic shock. Only mammalian meats pose this problem for sufferers of the allergy; meats such as beef, pork and venison are therefore off limits, but chicken, turkey and fish can all be eaten as they are not mammals and do not produce alpha-gal.

Unfortunately for any sufferers of the condition, there's no cure or fix for the allergic reaction, nor do doctors or scientists know if the reaction eventually wears off over time. As with other allergies, it's recommended that sufferers avoid red meat and carry an EpiPen to guard against anaphylactic shock. While the allergy is largely confined to the US states where the lone star tick is prevalent, cases have also been reported in France, Australia and several other countries due to bites from different species of tick, so it's an allergy that is considered to be on the rise.

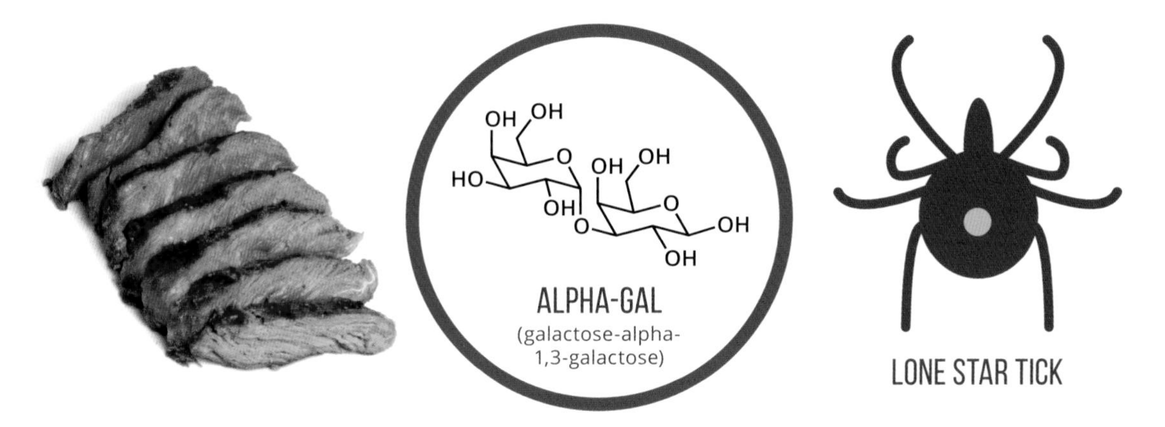

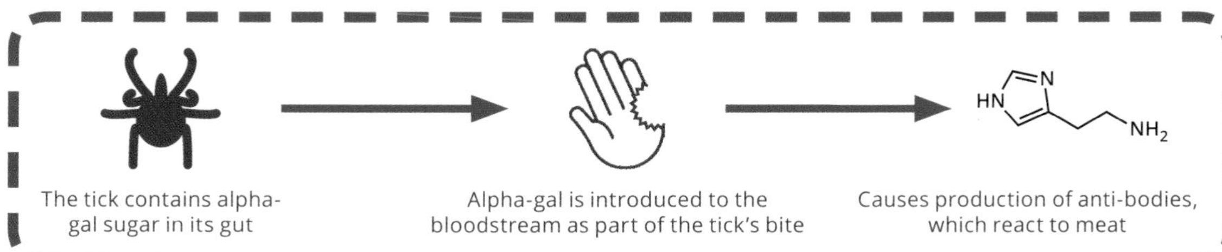

WHAT MEATS CAN A SUFFERER EAT?

CHICKEN

TURKEY

BEEF

PORK

VENISON

THE CHEMICAL COMPOSITION OF CLOVE OIL

HO
O

EUGENOL
70–85% of essential oil

O
O
O

EUGENYL ACETATE
15% of essential oil

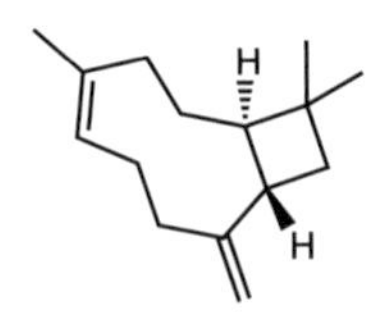

BETA-CARYOPHYLLENE
5–10% of essential oil

THE PROPERTIES OF EUGENOL

Eugenol has antiseptic, anti-inflammatory, antifungal and analgesic properties, though the efficacy of its use combatting dental pain is debated.

THE AROMA OF CLOVES

HO
O

EUGENOL
woody, spicy odour

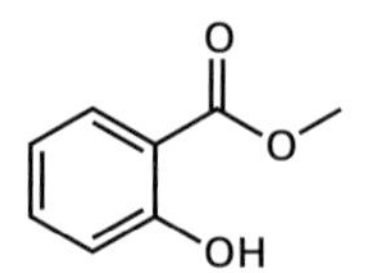

METHYL SALICYLATE
wintergreen/minty odour

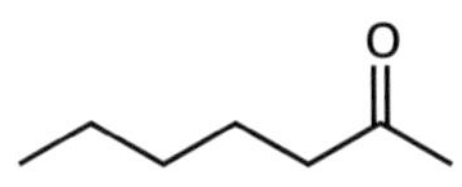

2-HEPTANONE
fruity, spicy odour

WHY CAN CLOVE OIL BE USED AS AN ANTISEPTIC?

Cloves are a spice that you may well have stowed away somewhere in the kitchen. Originally from the Maluku Islands in Indonesia, the spice itself comes from the flower buds of the clove tree. It is used to flavour food, imparting a sweet, aromatic flavour, and is commonly one of the spices used in the making of mulled wine. More than this, the oil of cloves is traditionally recommended as a remedy for relieving toothache, amongst other conditions. So, what are the chemicals that make this use possible?

To start with, let's consider the composition of clove oil. There are actually three types, depending on whether the oil is extracted from the buds, the leaves or the stems of the clove tree; here, we'll focus mainly on the bud oil. This is composed of three main compounds, as well as myriad minor constituents: eugenol, which accounts for around 70–85 per cent, eugenyl acetate, which accounts for around 15 per cent, and beta-caryophyllene, which accounts for 5–10 per cent.

Eugenol is the major compound that allows clove oil to be used as a remedy for toothache. It has an impressively wide variety of properties: it's an anaesthetic and antiseptic, and has anti-inflammatory, antifungal, antibacterial and insecticidal properties – quite the catalogue. In terms of toothache, we're mainly concerned with the anaesthetic properties. These come about due to eugenol's ability to affect nerves in areas to which the clove oil is applied. It inhibits the movement of sodium ions, lessening the nerves' ability to communicate with the brain and transmit the sensation of pain.

Despite this, the FDA has stated that there isn't currently enough evidence to rate eugenol as effective for the treatment of toothache. This isn't to say it has no effect, and there are studies that have shown that it performs more effectively against toothache than a placebo, but that this effect has not been shown to be significant enough to be of major use.

Another, odder, use for clove oil is as a topical cream, in combination with some other ingredients, targeted at preventing premature ejaculation. It's likely that this use also exploits eugenol's effect on nerves.

Finally, the aroma of cloves is influenced by the presence of eugenol, but also by the presence of some minor compounds in the composition. One of these is methyl salicylate, an ester commonly referred to as oil of wintergreen; another is 2-heptanone, which has a fruity, spicy odour. 2-heptanone is particularly interesting; much like eugenol, it can act as an anaesthetic, and research has shown that it is also contained in the mandibles of honeybees. The compound is secreted when the honeybee bites intruders in its hive, paralysing the intruder and allowing it to be removed by the bee. This is a comparatively recent discovery, and the compound has been patented for potential use as an anaesthetic in humans in the future.

DOES MSG CAUSE 'CHINESE RESTAURANT SYNDROME'?

Monosodium glutamate, or MSG for short, has long been the villain of the food supplement world. In the UK, Chinese takeaways proudly display 'No MSG' signs beside their counters, and many websites will purport to tell you 'the truth about MSG'. The real truth about MSG is that it's the victim of a character assassination of the highest order – as a short examination of the history of and research on MSG will reveal.

MSG was first isolated from seaweed in Japan in 1908. It was said to contribute an umami flavour when added to dishes; 'umami' is derived from the Japanese word for tasty. By the mid-twentieth century, MSG was a commonly used supplement in Japanese and Chinese cuisine, and had also spread to numerous other countries, including the USA, where it was routinely used in restaurants and takeaways across the country.

The term 'Chinese restaurant syndrome' was coined by a Chinese–American doctor, Robert Ho Man Kwok, who wrote a letter to a scientific journal complaining of experiencing palpitations and numbness after eating in Chinese restaurants. Kwok didn't identify any particular component of his meal as causing this effect but, despite the scarcity of evidence, MSG was quickly fingered as the culprit. A study carried out by a Dr John Olney around the same time found that when MSG was injected into the brains of mice, it could cause brain damage.

While this may seem concerning, an oft-omitted fact when reporting this study is that Olney used huge quantities of MSG in his studies, up to four grams per kilogram of body weight all at once, an amount many times higher than humans are likely to consume as part of a balanced diet. To put this into perspective, in industrialised nations it's estimated we ingest no more than one gram over the course of a day. In order to match the highest dosage used in Olney's tests, we'd have to consume 300 grams of MSG all at once – a quantity many times more than the amount of MSG found in an average Chinese takeaway meal.

A study that gets swept under the carpet is one carried out in the 1970s, which for six weeks fed 11 subjects up to almost 150 grams of MSG, and noted no ill effects as a consequence. The fact of the matter is that, despite the plethora of symptoms that MSG has been linked to over the years, there is absolutely no scientific evidence for any of them. Numerous studies and reviews have failed to find any correlation between undesirable symptoms and MSG, and its use as a food supplement is still approved by food regulatory bodies.

Chemically, MSG is simply the sodium salt of glutamic acid, which is a naturally occurring amino acid. Glutamic acid is found in tomatoes, ham and cheese, and is chemically the same as MSG – both are treated in exactly the same way by the body. If MSG did cause the symptoms commonly attributed to it, then you'd fully expect eating foods high in glutamic acid to produce exactly the same effect. Oddly, you don't tend to hear anyone complaining of 'Chinese restaurant syndrome' after eating cheese.

GLUTAMIC ACID
naturally occurring amino acid

MONOSODIUM GLUTAMATE
sodium salt of glutamic acid

FOODS WHICH NATURALLY CONTAIN GLUTAMIC ACID

A large number of studies have failed to link MSG to the symptoms commonly described as 'Chinese restaurant syndrome'. There is no evidence that MSG is harmful to humans at normal dietary levels.

THE 'SYMPTOMS' OF 'CHINESE RESTAURANT SYNDROME'

HEADACHE

SWEATING

NAUSEA

FATIGUE

SUCROSE

ASPARTAME

SUCRALOSE

COMPARISON OF THE SWEETNESS OF ARTIFICIAL SWEETENERS VS SUGAR (SUCROSE)

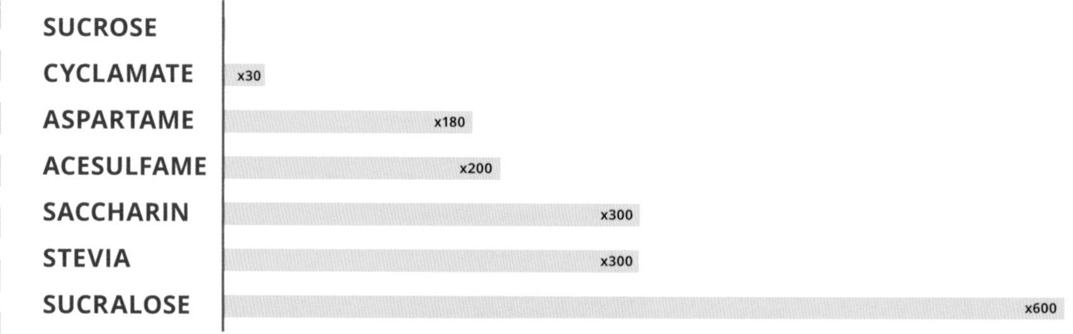

The artificial sweetener lugduname is estimated to be as much as 300,000 times sweeter than sugar, but is not yet approved for use in food.

WHY ARE SWEETENERS SOMETIMES USED IN PLACE OF SUGAR?

Sugar makes foods taste great, but we all know that too much of it can be a bad thing. It plays a major role in tooth decay – the only disease towards which sugar has a scientifically proven contribution. Plaque-forming bacteria can utilise sugar as energy in order to boost their growth, hence the constant warnings to maintain a sensible sugar intake in your diet.

Granulated sugar as we commonly encounter it is known chemically as sucrose. We might think of all natural sugars as being the same, but in fact there is a large range. The sugar commonly found in fruits is fructose. Glucose is another common sugar, and the one synthesised by plants during photosynthesis. In fact, sucrose is just a molecule of glucose and a molecule of fructose bonded together. Natural sugars can have varying sweetnesses – for example, fructose has a slightly sweeter taste than sucrose, whereas the sugar found in milk, lactose, is less than half as sweet as sucrose.

The risk with all natural sugars is the spectre of tooth decay, which is one of the many reasons why a lot of manufacturers instead turn to the use of artificial sweeteners. There are a wide number of different sweeteners approved for use – aspartame, saccharine, sucralose and stevia are some of the more widely known. Although these compounds all have differing structures, there is a theory of why they all produce the sensation of sweetness.

The 'triangle theory of sweetness' suggests that a molecule must have three different groups of atoms present – a carbonyl group (C=O), an amide group (N-H), and a hydrophobic group. These groups must also be arranged in a particular manner in space. The carbonyl and amide must be approximately 0.3 nanometres apart, while the hydrophobic group must be situated further away from the amide group than the carbonyl group. While not absolutely every sweet compound obeys these rules, it seems to be a relatively good approximation.

Artificial sweeteners are many times sweeter than sucrose. Aspartame and saccharin are both around 300 times sweeter, and others can be as much as 2,000 times sweeter. This means that much less is needed in order to produce the same level of sweetness as sucrose, saving manufacturers money. Additionally, artificial sweeteners don't promote tooth decay because plaque-forming bacteria cannot feed on them.

Despite these benefits, artificial sweeteners have something of a bad reputation. Aspartame, in particular, comes in for bad press and is accused of causing cancer. Reviews on the safety of aspartame makes it clear that no link with cancer or brain tumour has been found at concentrations of aspartame many times higher than that used in everyday products. A more recent study has suggested that some artificial sweeteners may induce glucose intolerance, a risk factor for diabetes. However, this study focused on tests of one sweetener, saccharin, in mice, along with very limited human studies. As such, its conclusions require further investigation before they can be confirmed.

WHAT ARE SULFITES AND WHY ARE THEY IN ALCOHOLIC DRINKS?

When drinking a bottle of wine or can of cider, you may have noticed that, somewhere on the can or bottle, there may be a short notice: 'contains sulfites'. You might well wonder what these compounds are, and why modern drinks manufacturers feel the need to add them into your drink. However, the thinking behind the use of sulfites as additives actually dates back many centuries.

The concept was supposedly first pioneered by the Romans – they discovered that by burning candles made of the element sulfur in empty wine bottles, the wine they were then filled with was less prone to going off and producing a vinegar-like taste and smell. The reason this practice was effective was because burning sulfur produces sulfur dioxide gas. Adding sulfur dioxide to wine later became common practice in the Middle Ages.

Sulfur dioxide has a wide range of preservative functions: firstly, it is an antioxidant, slowing the progress of oxidation reactions. In wines, this prevents the oxidation of some of the organic compounds in its composition, in turn helping to preserve the wine's colour and flavour. It's also an antimicrobial, particularly important in wine and other drinks to which it is added, as it inhibits the growth of yeasts and moulds, as well as the growth of food-poisoning bacteria.

These days, sulfur dioxide is often added directly to alcoholic drinks in the form of sulfites. The most frequently utilised compound is sodium metabisulfite, but potassium metabisulfite is also used. When mixed with water, these compounds produce sulfur dioxide, which can then go on to have the previously stated effects in the drink. Although we're discussing sulfites in the context of alcoholic drinks, they also have a wide range of applications in other foods – such as being used as a preservative in dried fruits and meat products.

In fact, the yeasts in wine and beer naturally produce some sulfur dioxide themselves, though the levels do need to be topped up artificially in order to reach the most effective concentration. So, there's seemingly little to be concerned about when it comes to sulfites in our food and drink. That said, some occasional sensitivity to sulfites amongst asthma sufferers has been noted. It's estimated that 3–10 per cent of asthma sufferers may show some reaction to the presence of sulfites; this reaction could range from very mild symptoms to a more serious asthma attack. For these people, the only solution is to avoid those products which are known to be high in sulfites.

$2K^+ \left[O_3S{-}SO_2 \right]^{2-}$

POTASSIUM METABISULFITE
preferred additive

$2Na^+ \left[O_3S{-}SO_2 \right]^{2-}$

SODIUM METABISULFITE
sometimes used as a substitute

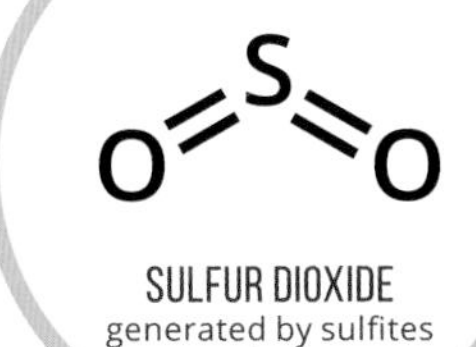

SULFUR DIOXIDE
generated by sulfites in drinks

THE PURPOSE OF SULFITES

Sulfites produce sulfur dioxide gas, which is useful for a number of reasons in a range of foods and drinks.

ANTIOXIDANT

Delays reactions that cause deterioration of food and decolourisation of foods and drinks.

ENZYME INHIBITION

Delays enzymatic reactions in foods; for example, those that lead to browning of fruits.

ANTIMICROBIAL

Restricts growth of moulds, yeasts and bacteria in both foods and drinks.

APPROXIMATE MAXIMUM SULFUR DIOXIDE CONTENT IN DIFFERENT FOODS & DRINKS

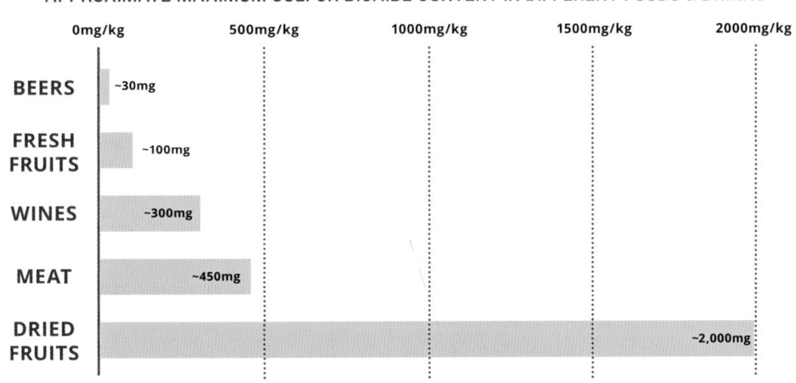

TRANSFORMATION

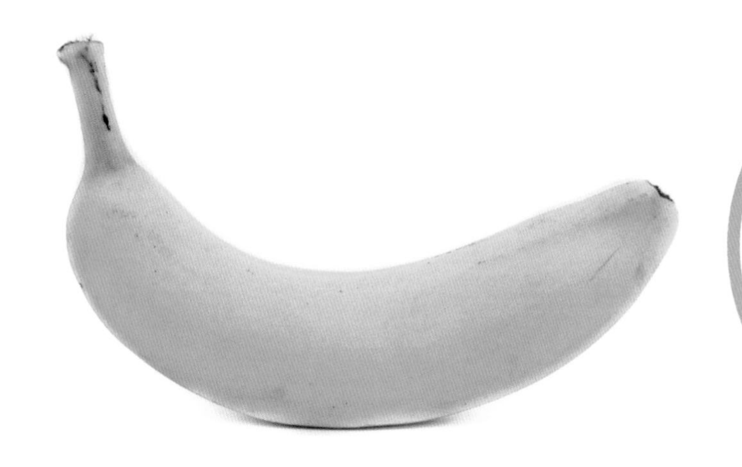

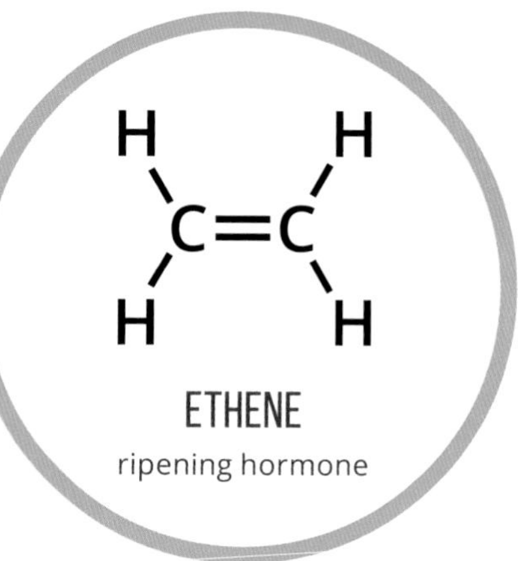

THE ENZYMES INVOLVED IN FRUIT RIPENING

PECTINASE
Breaks down plant cell walls – softening the fruit's flesh.

$C_6H_{12}O_6$

AMYLASE
Breaks down carbohydrates, forming the sugars that make fruit taste sweet.

HYDROLASE
Breaks down chlorophyll, causing the change in colouration associated with ripening.

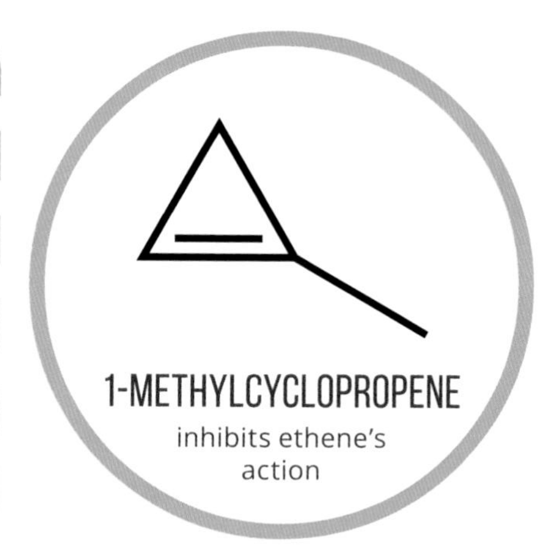

DO BANANAS HELP OTHER FRUIT TO RIPEN QUICKER?

Bananas are grown in 107 countries, and although at one time they may have been considered exotic, they're now one of the most popular fruits worldwide. Picked green from the trees, they slowly ripen to a yellow colour – but common advice is not to keep them in the fridge, otherwise their skin turns brown, or even black. This is due to the enzymatic browning reactions of the type we discussed when considering avocados. Leading on from that, a tactic commonly suggested when wanting to ripen avocados quicker is to put them in a plastic bag with a banana.

This may seem like a bit of an odd suggestion, but by doing this you're actually taking advantage of a particular chemical produced by bananas. The chemical in question is ethene, a deceptively simple-looking compound which functions as a ripening hormone in fruits. It works by 'switching off' particular genes in the fruit when it is produced, which eventually leads to other genes, responsible for making enzymes which help the ripening process, being 'switched on'. The ethene produced by bananas can quite easily stimulate this process in other fruits, too, hence the plastic bag suggestion.

This is a process that we use to our advantage. If bananas were picked when already yellow, they'd most likely be brown and beginning to rot by the time they reach our supermarket shelves. Therefore, they are picked from plantations when they are still green and unripe, and transported in this condition. To ensure that they don't over-ripen in transit, other chemical gases can be used to slow the ripening process. This is usually accomplished by the use of 1-methylcyclopropene, which is capable of blocking the effects of ethene on the fruit, arresting its ripening.

Once the fruits have been shipped, it's now likely that they'll still be some way off fully ripe. In these cases, artificial ripening can be used. Taking advantage of the effect of ethene on accelerating the ripening process, the fruit can be gassed with ethene, and over the course of a day or two will approach full ripeness.

Ethene has one other strange effect – it can also affect some flowering plants, so leaving a banana in the proximity of a flowering plant could make the blooms appear slightly quicker.

WHY DOES ADDING SOME FRUITS TO JELLY STOP IT SETTING?

If you've ever tried to make fruit jelly, you might have noticed that certain fruits make the process trickier than others. In particular, pineapple, papaya and kiwi fruit have the rather unhelpful trait of seemingly preventing the jelly from setting, no matter how long they are left. There's a chemical reason for this, but before we consider it, we also need to think about what's actually going on when jelly sets.

Jelly is made up of gelatin, which is itself a processed version of the protein collagen, commonly found in both humans and animals. When you add hot water to gelatin, the weak forces holding the strands of gelatin together are overcome, and the individual proteins are then free to move around. When you put the jelly in the fridge to set, the protein molecules once again tangle and intertwine as they cool, trapping water as they do so and producing the characteristic consistency and appearance of jelly.

Pineapple, papaya and kiwi fruit all interfere with this process due to the enzymes they contain. Pineapple contains and enzyme called bromelain, kiwi fruit an enzyme called actinidin, and papaya one called papain. While their names are different, their purpose is the same: all three are protein-digesting enzymes. It might seem a little strange that some fruits can contain enzymes to digest proteins; it's thought that their presence can act as protection for the fruit against parasites and bugs. When these enzymes are mixed with gelatin, they break up the gelatin protein into smaller sections. These smaller sections are then not long enough to form the tangled structure as they cool, meaning that the jelly will not set.

If you're still desperate to make pineapple jelly, the good news is that there's a way around this problem. If you used tinned pineapple instead of fresh the jelly will quite happily set. This is due to the fact that tinned pineapple will have been heated to kill any bacteria. The heating of the pineapple will break down or 'denature' the enzymes in the pineapple, including the protein-digesting enzymes that make it impossible to make jelly with fresh pineapple. Once these are no longer functioning, you can make jelly with pineapple as you would with any other fruit.

GELATIN

The chemical structure of gelatin is formed of long chains of amino acids, the building blocks of proteins. A typical structure of a section of gelatin is shown below.

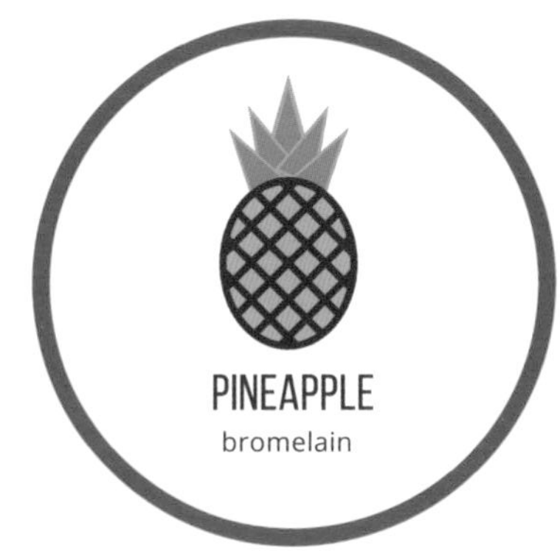

PINEAPPLE
bromelain

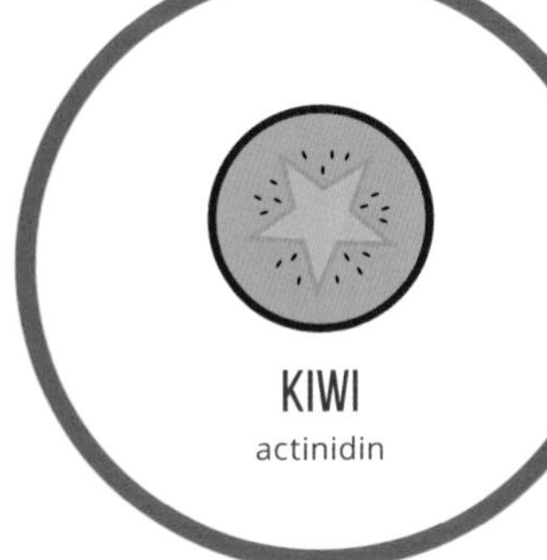

KIWI
actinidin

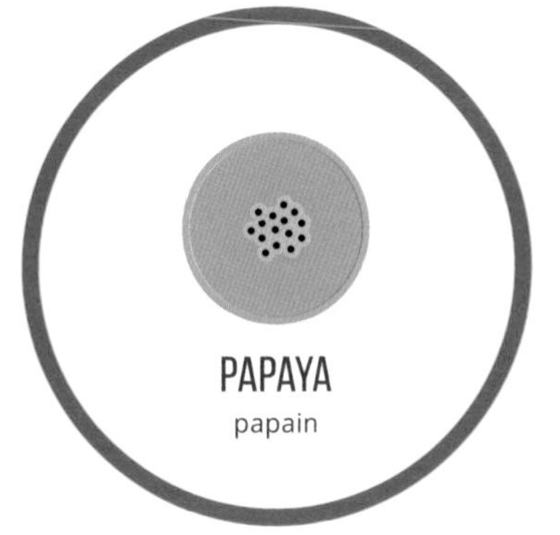

PAPAYA
papain

THE ENZYMES IN THESE FRUITS BREAK UP THE PROTEIN STRUCTURE OF GELATIN, PREVENTING IT FROM SETTING

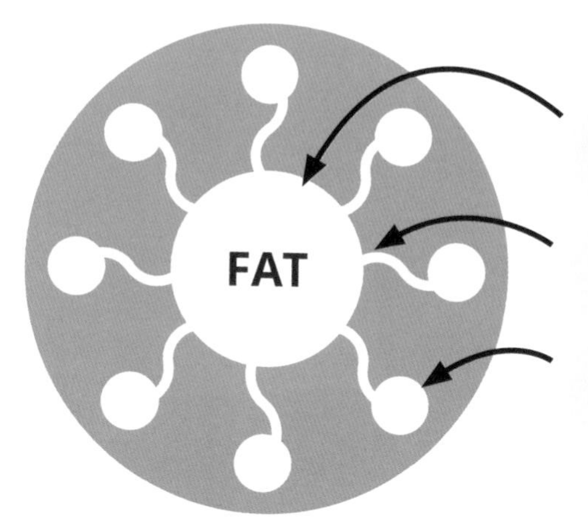

HOW THE CREAM EMULSION IS FORMED

Cream is an emulsion of fat droplets in water. Protein molecules in cream act as emulsifying agents.

One end of the protein molecules is 'hydrophobic'. It won't dissolve in water, but dissolves in fats.

The other end is 'hydrophilic'. It dissolves in water, allowing the formation of the emulsion.

WHAT HAPPENS WHEN CREAM IS WHIPPED?

air bubbles incorporated

structure of fat droplets disrupted

fat droplets link and form a gel structure

layers of water and fat, linked by proteins, cause cream to thicken

OTHER EXAMPLES OF **EMULSIONS**

PAINT

MAYONNAISE

BUTTER

MILK

WHY DOES WHIPPING CREAM MAKE IT THICKER?

Whipping cream is commonly served with desserts; as you're no doubt aware, it comes in relatively runny form if you purchase it in a carton rather than in one of the aerosol cans that does all of the hard work for you. Of course, before you serve it with dessert it has to be whipped in order to thicken it – but why does this cause the cream to thicken?

Let's start with the basics. Cream is what scientists call an emulsion; in everyday speak, this might make you automatically think of paint, and paint is a type of emulsion, but many other mixtures can be emulsions too. An emulsion is a mixture of two (or more) liquids, where one liquid is in the form of millions of microscopic droplets dispersed through the other. Cream is an emulsion of fat droplets in water.

To stop emulsions separating, often emulsifying agents are necessary. Emulsifying agents generally have two parts to their molecules. One end will dissolve in water, but not fats or oils – this is known as the hydrophilic, or 'water-loving' end of the molecule. The other end, which will dissolve in fats and oils, but not water, is known as the hydrophobic, or 'water-hating', end of the molecule. In cream, proteins act as the emulsifying agents, helping to stabilise the emulsion by surrounding the fat droplets and allowing them to remain dissolved in the water. Washing-up liquid is another example of an emulsifying agent – by helping water and oil to mix, it allows you to wash up your dishes effectively.

So far, then, we understand what cream is, but we're not any closer to understanding why it thickens when whipped. Let's think about what we're introducing when we whip cream. By whipping it over several minutes we're incorporating more and more air into the mixture. These tiny air bubbles which become trapped in the mixture can disrupt the fat droplets and cause them to link up together while still mixed in with the water. The end result is the creation of a foam, which can be viewed as layers of water and fat, linked by the proteins between them, with microscopic bubbles of air also included. This leads to the marked change in texture and appearance of the cream.

SHOULD YOU KEEP CHOCOLATE IN THE FRIDGE?

Whether or not to store chocolate in the fridge is, amongst some, a controversial and divisive issue. However, the chemistry behind chocolate's structure may just help settle the argument.

Cocoa butter is the primary ingredient in chocolate; it is composed primarily of fat molecules, and how they are arranged determines the structure of chocolate. The molecules themselves don't change at all in these different structures; what does change is how they are arranged or stacked together.

The ability of a substance to display different structures is known as polymorphism, and cocoa butter has at least six different structures, or crystal forms. These differ in how the molecules are arranged, which in turn influences their properties, such as appearance, taste and texture. This can affect the taste and quality of the chocolate.

The best crystal form for chocolate, in terms of it exhibiting the best appearance and taste, is form V. This form has a shiny appearance, produces an audible snap when broken, melts in the mouth and has a smooth texture. The other forms can be soft and crumbly, and often exhibit sugar or fat bloom. Sugar bloom is caused by the introduction of moisture to the chocolate, causing some of the sugar molecules to be drawn out. Fat bloom is the result of partial melting causing fats to rise to the surface of the chocolate.

Form V might be the most desirable chocolate polymorph, but unfortunately it's not the most stable of the six, and requires a process called tempering in order for it to be the main structure present. If melted cocoa butter is simply allowed to cool naturally, a mixture of forms I–V will be obtained. Tempering involves allowing the melted chocolate to cool very slowly, as this increases the amount of form V that is formed in the mixture. Once cooled, the chocolate can then be heated again to just below the melting point of form V – this melts forms I–IV, which all have lower melting points, but not the form V crystals. When the chocolate is again allowed to cool, it solidifies following the pattern of the existing form V crystals, with the end result being chocolate that has very little of the other forms of structure present.

Form VI doesn't form as melted chocolate solidifies. Instead, it forms only after several months, from form V. The fat molecules in form V have enough energy over this time period to change to form VI, which is harder, and melts slowly in the mouth due to the higher melting point. It's also possible to see fat bloom forming on the chocolate.

The transformation of form V to form VI can be arrested simply by storing the chocolate in the fridge. This is because, at a lower temperature, the molecules in the structure don't have enough energy to convert to form VI.

Does this settle the chocolate in the fridge argument? Well, not quite. Sudden changes in temperature can also have a negative impact on the structure of the chocolate, and so putting chocolate straight into the fridge on a hot day could well prove just as detrimental to its quality. So, if the thought of putting chocolate in the fridge is an anathema to you, there is a justification you can cling to.

The molecules in cocoa butter can be stacked together in different ways – these differing forms are known as **polymorphs**. Each polymorph has differing properties, which can impact on the quality of the chocolate. The most desirable is form V.

FORM	I	II	III	IV	V	VI
MELTING POINT	17.3˚C	23.3˚C	25.5˚C	27.3˚C	33.8˚C	36.3˚C
SOFT	✔	✔				
FIRM			✔	✔	✔	
HARD						✔
CRUMBLY	✔	✔				
GOOD 'SNAP'					✔	
SHINY					✔	
FAT 'BLOOM'	✔	✔	✔	✔		✔

INCREASING STABILITY & DENSITY

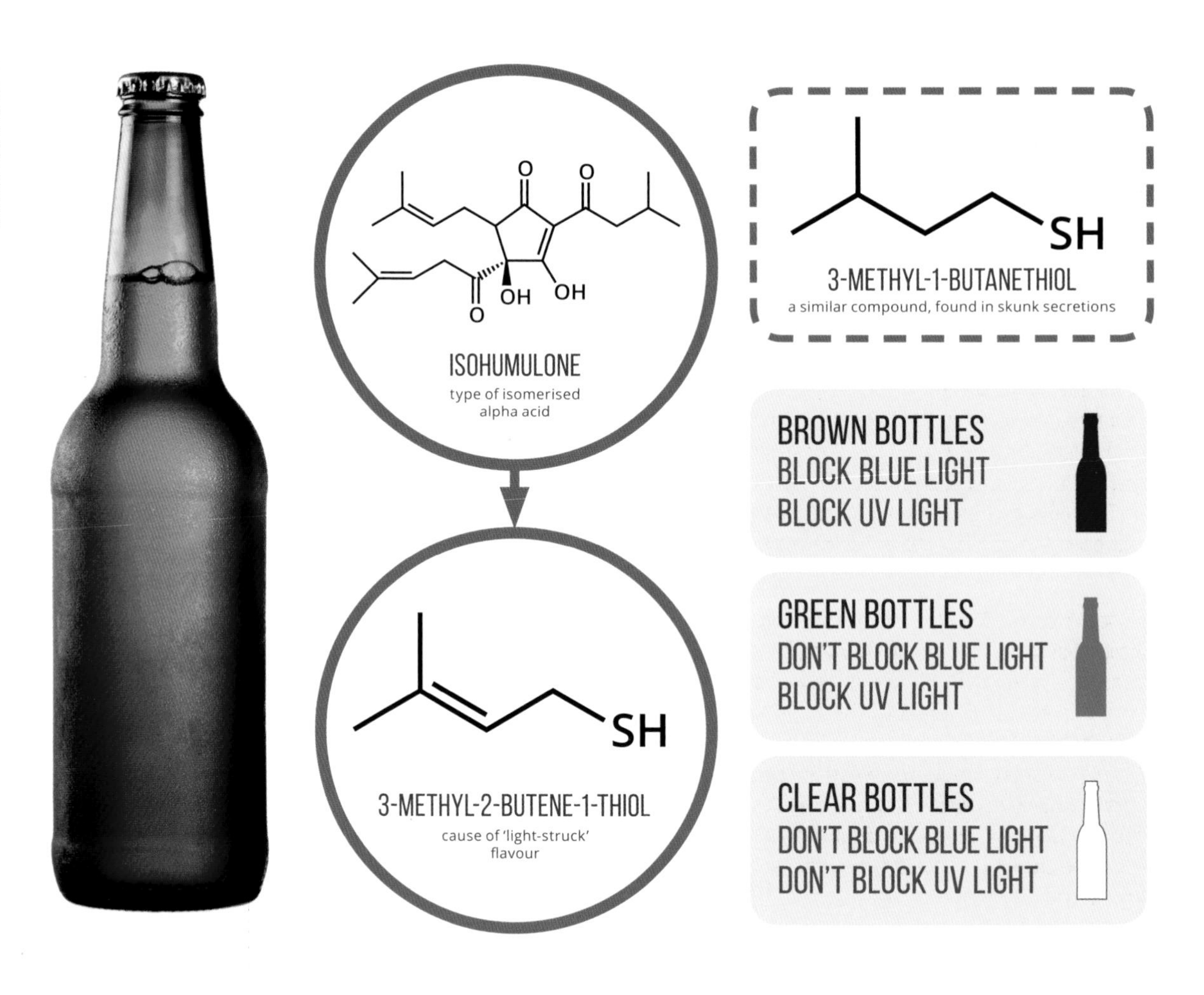
O
O
OH
OH
O
ISOHUMULONE
type of isomerised alpha acid
SH
3-METHYL-1-BUTANETHIOL
a similar compound, found in skunk secretions
SH
3-METHYL-2-BUTENE-1-THIOL
cause of 'light-struck' flavour
BROWN BOTTLES
BLOCK BLUE LIGHT
BLOCK UV LIGHT
GREEN BOTTLES
DON'T BLOCK BLUE LIGHT
BLOCK UV LIGHT
CLEAR BOTTLES
DON'T BLOCK BLUE LIGHT
DON'T BLOCK UV LIGHT

WHY ARE BEER BOTTLES USUALLY MADE OF DARKENED GLASS?

You may have noticed that beer in supermarkets is almost exclusively sold in bottles made of darkened glass or opaque cans. This isn't merely a question of aesthetics – there's scientific reasoning behind it. You might not think of beer as something that can easily go off, but the darkened glass bottles or cans are essential to prevent a process that accomplishes exactly that.

Hops are used in the brewing process, and are an essential contributor to the flavour and bitterness of beers. During the brewing process the alpha acids in the hops degrade to form a very slightly different set of compounds, isomerised alpha acids. These compounds add a large proportion of the bitter taste to beer; they're also important in the process of beer going off.

When beer is exposed to light, the photons of light kick off a process that causes the reaction of some of the isomerised alpha acids with another compound found in beer, riboflavin, which produces a compound called 3-methyl-2-butene-1-thiol, or MBT for short. MBT isn't too dissimilar in structure from some of the compounds found in skunk spray, and has a similarly unpleasant odour. For this reason, this process is often referred to as 'skunking', and due to the involvement of light in the process, beer affected by it is commonly known as 'light struck'.

This is where the darkened glass bottles come in. It's not the whole spectrum of light that kick-starts the skunking reaction, but rather specific portions of it. The regions responsible are light at the blue end of the spectrum, with wavelengths between 400 to 500 nanometres, and ultraviolet light, with a wavelength below 400 nanometres. Brown beer bottles block both of these wavelengths, and cans, being opaque, obviously do likewise, making either of these the best choice for avoiding light-struck beer. Green beer bottles block the ultraviolet portion of the spectrum, but not the light at the blue end, meaning that beer contained within green bottles is a little more susceptible to skunking. In some beers, a small amount of skunking is actually considered to be part of the flavour, though obviously too much can still be a bad thing.

Some beers seem to fly in the face of the logic of preventing skunking by being sold in clear bottles. This is because they use very small quantities of hops, hence minimising the isomerised alpha acid content of the beer. Other brewers may use a special hop extract, tetra-hop, to provide bitterness; although this still contains isomerised alpha acids, their molecular structure is marginally different from those contributed by natural hops, and so it avoids the light-struck phenomenon. More commonly, though, the colour of the bottle is a question of marketing and serving suggestions, such as drinking the beer from the bottle with a slice of lime, are designed to minimise the impact of any skunk-like taste that develops.

WHAT CHEMICAL COMPOUNDS MAKE JAMS SET?

If you've ever tried your hand at jam-making, you'll know that it's something of a tricky process. A number of factors need to be just right to achieve a perfectly set jam – and chemistry can help explain why. There are three key chemical entities that go into jam-making: sugar, pectin and acids. Here, we'll look at each in turn, and how they help jam achieve its eventual consistency.

PECTINS Pectins are long, linked chains of sugar molecules, which are found naturally in plant cell walls. Although we refer to them in general as 'pectin', their structures are variable, as well as hard to determine; a rough general structure is given in the graphic, but in reality the overall structure can be much more complicated. Pectins are found in fruits, particularly in the peels and cores. When jam sets, pectin plays a vital role.

Boiling the jam releases the pectins from the fruit used; with the correct amount of sugar and acidity, the long pectin chains can bind to each other via intermolecular interactions, forming a gel network. This network generally forms at the 'setting point' of jam, which is approximately 104°C. Once it has formed, the jam can be allowed to cool, and the gel network 'traps' the water content of the jam, leading to setting.

SUGAR An important part of jam is, of course, the sugar content, which is vital for the flavour and also plays a role in helping jam set. Many jam recipes recommend the use of a 1:1 ratio of fruit to sugar in jam-making. As well as sweetening the jam, the sugar also helps the pectin set – it enhances the pectin's gel-forming capability by drawing water to itself, decreasing the ability of the pectin to remain in separate chains. Additionally, sugar imparts a preservative effect. By binding water molecules to itself, it reduces the amount of water available in the jam, to the point at which it is too low for microbial growth, helping to ensure that the jam doesn't go off too rapidly after it's been made! The final sugar content of jam should be between 65 and 69 per cent.

ACIDS Acids are also important in helping the pectin to set. The COOH groups in the pectin are usually ionised, and the negative charges on the molecules this ionisation causes can cause repulsion, and prevent the pectin chains from being able to form the gel network. To avoid this, we need the pH of the mixture to be roughly in the range of 2.8–3.3. At these more acidic pHs, the COOH groups aren't ionised, lowering the magnitude of the repulsive forces.

Fruits naturally contain acids – the most well known is citric acid, but malic acid and tartaric acid are also found in a number of fruits. Whilst some acid will be contributed by the fruit from which the jam is made, often this won't be enough to reach the desired pH, and for this reason more must be added. This is commonly in the form of lemon juice, which contains citric acid, though powdered forms of acids can also be used.

In summation, then, the three factors of pectin, sugar and acid have to be in perfect balance for jam to set. If it doesn't, you can often point to one of those three factors being somehow amiss – and understanding the chemistry behind why jam sets in the first place can often help you identify how to fix it!

SUGAR

SUCROSE
(table sugar)

65–69%

REQUIRED FINAL SUGAR CONTENT OF JAM

SETTING & PECTINS

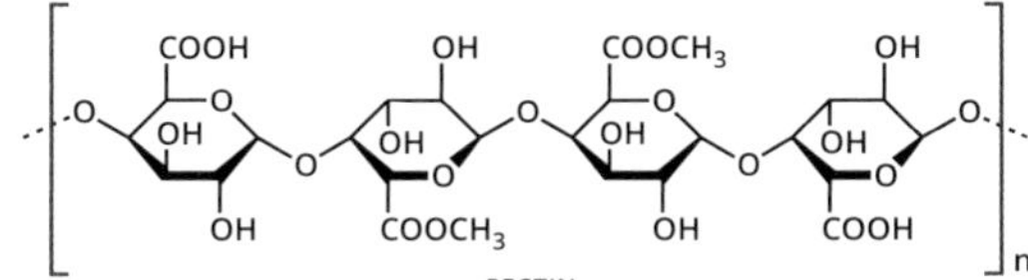

PECTIN
(typical chemical structure)

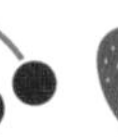

LOW IN PECTIN
Pears, peaches, cherries, strawberries, raspberries, blackberries, sweet plums, blueberries, elderberries.

HIGH IN PECTIN
Apples, gooseberries, blackcurrants, sour plums, grapes, citrus rind.

pH

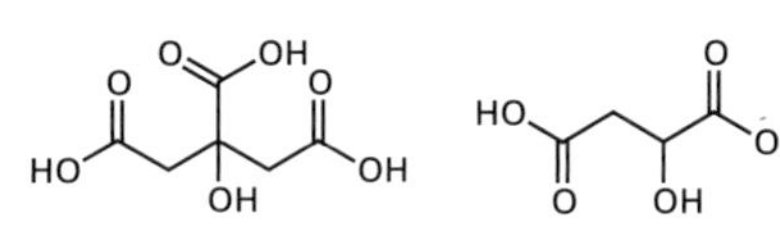

CITRIC ACID
(occurs naturally in citrus fruits)

MALIC ACID
(occurs naturally in apples)

2.8–3.3

OPTIMAL pH FOR SETTING

WHAT CAUSES THE BITTERNESS AND DRY SENSATION IN RED WINE?

Bitterness and astringency are characteristic of a large number of red wines. Bitterness needs no explanation, whilst astringency is commonly described as a drying, puckering sensation in the mouth. In general, red wine is a complex mix of a large number of chemicals; there's no exact figure, but estimates range from around 800 different compounds to over 1000. Compounds referred to as flavonoids comprise just 0.1 per cent of your average red wine – but these are the ones that cause the taste sensations.

A family of compounds called the flavan-3-ols contribute to the bitterness of wine. They originate primarily from the seeds of the grapes, with catechin and epicatechin the primary flavan-3-ols found in red wine; these compounds are also found in high concentrations in tea and dark chocolate, and have been associated with health benefits due to antioxidant activity.

The similar sounding flavonols also have a similar looking structure to the flavan-3-ols. However, the differences are significant enough that flavonols don't contribute to the bitterness of the wine, as the flavan-3-ols do – in fact, they've yet to have any sensory impact attributed to them. They, too, have antioxidant properties, but research suggests they're present in red wine in too low a concentration to be considered a good source. They do, however, help contribute to the colour of red wine.

The other family of compounds important to bitterness and astringency are the tannins. Tannins are polymers – that is, many smaller molecules joined together to make a long chain. More common examples of polymers are man-made plastics, or the cellulose in plants. Condensed tannins are the main class found in red wines, which consist of many different flavan-3-ol molecules joined together – as many as 27 in one polymer molecule when the grapes used to make the wine are first harvested. Some tannins can also come from the barrels in which the wine is aged.

The tannins in red wine contribute to its dryness, as well as the bitterness. When you drink wine, the tannins react with the proteins in your saliva. This forms a precipitate, and leads to the sensation of dryness. Obviously, variation of tannin concentration will affect the amount of dryness that is perceived. They can also contribute to the colour by combining with the anthocyanins. Tannins have also been implicated in the changes that occur as wine ages, though these are complex chemical processes and are yet to be fully understood.

Finally on the subject of tannins, they may also be the reason that some people experience headaches or migraines after drinking red wine. This has been shown to be a demonstrable effect of red wine in some people, and it was suggested that tannins could cause this by altering the levels of serotonin, a neurotransmitter, in the brain. However, the jury is very much out – a number of other possibilities have been suggested, but we're currently no closer to being able to pinpoint a specific molecule.

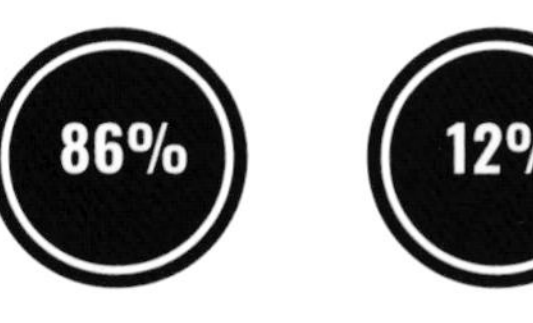

WATER

ETHANOL

GLYCEROL

ORGANIC ACIDS

0.1%

TANNINS & PHENOLICS

OTHER COMPOUNDS

NOTE THAT THESE FIGURES ARE FOR AN AVERAGE COMPOSITION – EXACT PERCENTAGES VARY DEPENDING ON THE PARTICULAR WINE

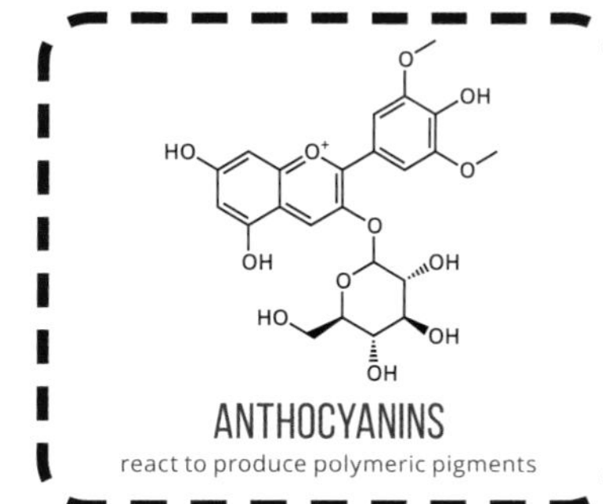

ANTHOCYANINS

react to produce polymeric pigments

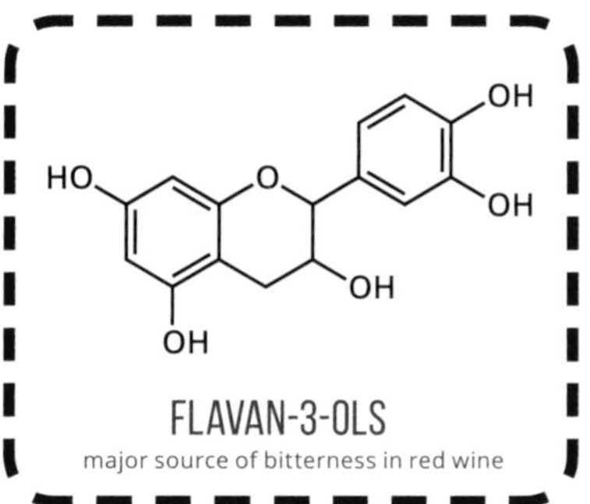

FLAVAN-3-OLS

major source of bitterness in red wine

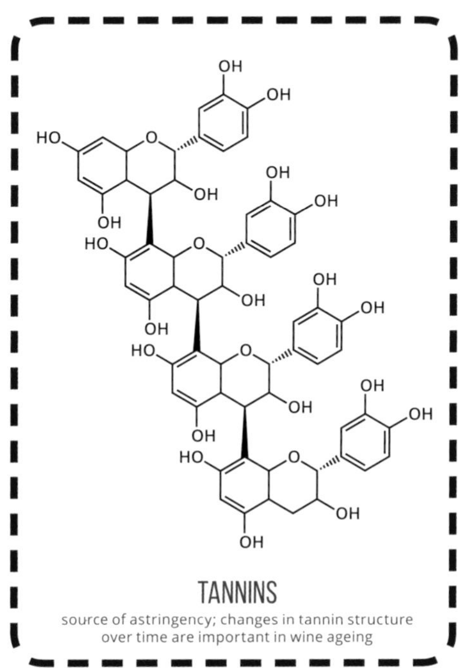

TANNINS

source of astringency; changes in tannin structure over time are important in wine ageing

As the bubbles in champagne rise to the surface, they carry flavour and aroma compounds with them; when they burst at the surface, the compounds are dispersed in fine liquid droplets, with some being significant contributors to champagne's aroma. A selection of compounds found within these bubbles are shown here – there are also many other compounds in the wine itself that contribute to flavour.

GAMMA-DECALACTONE

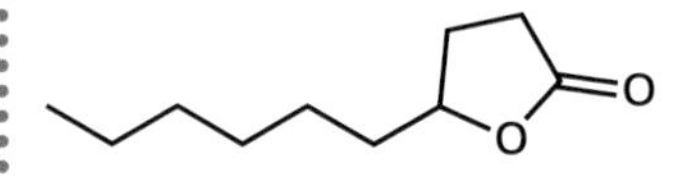

Fruity, peachy and sweet aroma

DECANOIC ACID

O
OH

Acid and toasty aromas

METHYL DIHYDROJASMONATE

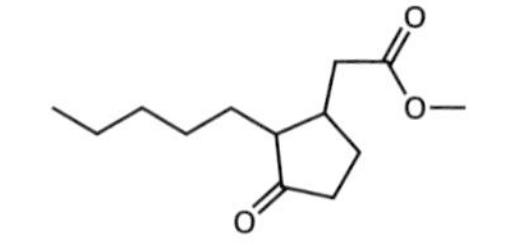

Sweet, fruity, floral aroma

7,8-DIHYDROVOMIFOLIOL

O
OH
OH

Contributor to fruity aroma

DODECANOIC ACID

O
OH

Dry and metallic notes

ETHYL MYRISTATE

O
O

Sweet and waxy aroma

HOW DO BUBBLES ENHANCE THE TASTE OF CHAMPAGNE?

Celebrations often call for the popping open of a bottle of champagne. The bubbles in the glass may seem simple enough, but there's actually a wealth of interesting chemistry behind them – chemistry that's vital for the perceived taste and aroma of the wine. There's a lot more to the bubbles than you might think.

The obvious chemical contributor that causes the bubbles to appear in champagne in the first place is carbon dioxide, which originates from the fermentation process. Champagne is unusual amongst wines, in that it undergoes two fermentations – one before bottling, and one in the bottle before it is drunk. The second fermentation produces the carbon dioxide and ethanol that are vital for the finished product.

An average 0.75 litre bottle of champagne contains around 7.5 grams of dissolved carbon dioxide – this may not sound like a lot, but when the bottle is opened, it would release around 5 litres of carbon dioxide gas if you allowed it to bubble until flat. In an individual champagne flute, assuming a volume of around 0.1 litres, approximately 20 million bubbles will be given off. This isn't even the bulk of the carbon dioxide – only around 20 per cent of it escapes from the wine in the form of bubbles, with the other 80 per cent escaping via direct diffusion.

As well as giving champagne its characteristic fizz, studies have shown that the bubbles are also vital contributors to the flavour and aroma of the wine. They can pull some compounds in the wine with them as they rise; when they reach the surface and burst, these compounds can be thrown into the air within tiny liquid droplets. Scientists have analysed the composition of these droplets, collected by holding a microscope slide over a champagne glass then transferring them to a solution, which was then run through a spectrometer to identify compounds present.

A large number of flavour and aroma compounds were discovered in the droplets, a selection of which are shown in the graphic. Hundreds of components were present, with some still yet to be identified, but interestingly, the composition of these droplets differs from that of the main body of the drink. This is due to the fact that only certain molecules are pulled up to the surface by the bubbles, influencing the droplet composition. The study's authors state that many of these compounds contribute to the aroma of the champagne, and the droplets dispersed by the bursting of the bubbles are therefore vital for both the aroma and flavour. Cheers to that!

REFERENCES

p.15 **BRUSSELS SPROUTS**

Drewnowski A, Henderson SA, Shore AB, Barratt-Fornell A. 1998. Sensory Responses to 6-n-Propylthiouracil (PROP) or Sucrose Solutions and Food Preferences in Young Womena. *Annals of the New York Academy of Sciences*. 855(1):797-801.

Turnbull B, Matisoo-Smith E. 2002. Taste sensitivity to 6-n-propylthiouracil predicts acceptance of bitter-tasting spinach in 3–6-y-old children. *The American Journal of Clinical Nutrition*. 76(5):1101-1105.

Drewnowski A, Henderson SA, Barratt-Fornell A. 2001. Genetic taste markers and food preferences. *Drug Metabolism & Disposition*. 29(4):535-538.

p.16 **ARTICHOKE**

Bartoshuk LM, Lee CH, Scarpellino R. 1972. Sweet taste of water induced by artichoke (*Cynara scolymus*). *Science*. 178(4064): 988-990.

p.19 **MIRACLE BERRIES**

Brouwer JN, Van Der Wel H, Francke A, Henning GJ. 1968. Miraculin, the sweetness-inducing protein from miracle fruit. *Nature*. 220:373-374.

Hiwasa-Tanase K, Hirai T, Kato K, Duhita N, Ezura H. 2012. From miracle fruit to transgenic tomato: mass production of the taste-modifying protein miraculin in transgenic plants. Plant cell reports. 31(3):513-525.

p.20 **ORANGE JUICE**

Allison A, Marie A, Chambers DH. 2005. Effects of residual toothpaste flavor on flavor profiles of common foods and beverages. *Journal of Sensory Studies*. 20(2):167-186.

p.23 **SMOKED MEAT**

Hindi SS. 2011. Evaluation of Guaiacol and syringol emission upon wood pyrolysis for some fast-growing species. *International Science Index*. 5(8):533-537.

Simon R, de la Calle B, Palme S, Meier D, Anklam E. 2005. Composition and analysis of liquid smoke flavouring primary products. *Journal of Separation Science*. 28(9-10):871-882.

p.24 **MILK**

Liu J, Yu CQ, Li JZ, Yan JX. 2001. Study on the deteriorating course of fresh milk by laser-induced fluorescence spectra. *Guang pu xue yu guang pu fen xi*. 21(6):769-771.

Bassette R, Fung DY, Mantha VR, Marth EH. 1986. Off-flavors in milk. *Critical Reviews in Food Science & Nutrition*. 24(1):1-52.

p.27 **CORIANDER**

Bhuiyan MNI, Begum J, Sultana M. 2009. Chemical composition of leaf and seed essential oil of *Coriandrum sativum* L. from Bangladesh. *Bangladesh Journal of Pharmacology*. 4(2):150-153.

Eriksson N, Wu S, Do CB, Kiefer AK, Tung JY, Mountain JL, Francke U. 2012. A genetic variant near olfactory receptor genes influences cilantro preference. *Flavour*. 1(1):22.

p.28 **DILL AND SPEARMINT**

Zawirska-Wojtasiak R. 2006. Chirality and the nature of food authenticity of aroma. Acta Sci Pol Technol Aliment. 5(1):21-36.

Fabro S, Smith RL, Williams RT. 1967. Toxicity and teratogenicity of optical isomers of thalidomide. *Nature*. 215:296.

p.31 **COFFEE**

Farah A, de Paulis T, Trugo LC, Martin PR. 2005. Effect of roasting on the formation of chlorogenic acid lactones in coffee. *Journal of Agricultural and Food Chemistry*. 53(5): 1505-1513.

Blumberg S, Frank O, Hofmann T. 2010. Quantitative studies on the influence of the bean roasting parameters and hot water percolation on the concentrations of bitter compounds in coffee brew. *Journal of Agricultural and Food Chemistry*. 58(6):3720-3728.

p.32 BEER

De Keukeleire D. 2000. Fundamentals of beer and hop chemistry. *Quimica nova.* 23(1):108-112.

Stevens R. 1967. The chemistry of hop constituents. *Chemical Reviews.* 67(1):19-71.

p.36 GARLIC

Cai XJ, Block E, Uden PC, Quimby BD, Sullivan JJ. 1995. Allium Chemistry: Identification of Natural Abundance Organoselenium Compounds in Human Breath after Ingestion of Garlic Using Gas Chromatography with Atomic Emission Detection. *Journal of Agricultural and Food Chemistry.* 43(7):1751-1753.

Munch R, Barringer SA. 2014. Deodorization of garlic breath volatiles by food and food components. *Journal of Food Science.* 79(4):526-533.

Lu X, Rasco BA, Jabal JMF, Aston DE, Lin M, Konkel ME. 2011. Investigating Antibacterial Effects of Garlic (*Allium sativum*) Concentrate and Garlic-Derived Organosulfur Compounds on Campylobacter jejuni by Using Fourier Transform Infrared Spectroscopy, Raman Spectroscopy, and Electron Microscopy. *Applied & Environmental Microbiology.* 77(15):5257-5269.

p.39 ASPARAGUS

Mitchell, SC. 2001. Food Idiosyncrasies: Beetroot & Asparagus. *Drug Metabolism & Disposition.* 29(2):539-543.

Pelchat ML, Bykowski C, Duke FF, Reed DR. 2011. Excretion and Perception of a Characteristic Odor in Urine after Asparagus Ingestion: a Psychophysical and Genetic Study. *Chem Senses.* 36:9-17.

p.40 DURIAN FRUIT

Li JX, Schieberle P, Steinhaus M. 2012. Characterization of the Major Odor-Active Compounds in Thai Durian (Durio zibethinus L.'Monthong') by Aroma Extract Dilution Analysis and Headspace Gas Chromatography–Olfactometry. *Journal of Agricultural and Food Chemistry.* 60(45):11253-11262.

Maninang JS, Lizada MCC, Gemma H. 2009. Inhibition of aldehyde dehydrogenase enzyme by Durian (Durio zibethinus) fruit extract. *Food Chemistry.* 117(2):352-355.

p.43 BACON

Timón ML, Carrapiso AI, Jurado Á, van de Lagemaat J. 2004. A study of the aroma of fried bacon and fried pork loin. *Journal of the Science of Food and Agriculture.* 84(8):825-831.

p.44 FISH

Mitchell SC, Smith RL. 2001. Trimethylaminuria: the fish malodor syndrome. *Drug Metabolism & Disposition.* 29(4):517-521.

Dyer WJ, Mounsey YA. 1945. Amines in Fish Muscle: II. Development of Trimethylamine and Other Amines. *Journal of the Fisheries Board of Canada.* 6(5):359-367.

p.47 BLUE CHEESE

Qian M, Nelson C, Bloomer S. 2002. Evaluation of fat-derived aroma compounds in Blue cheese by dynamic headspace GC/olfactometry-MS. *Journal of the American Oil Chemists' Society.* 79(7):663-667.

Dartey CK, Kinsella JE. 1971. Rate of formation of methyl ketones during blue cheese ripening. *Journal of Agricultural and Food Chemistry.* 19(4):771-774.

Lawlor JB, Delahunty CM, Wilkinson MG, Sheehan J. 2001. Relationships between the sensory characteristics, neutral volatile composition and gross composition of ten cheese varieties. *Le Lait.* 81(4):487-507.

Day EA, Anderson DF. 1965. Cheese Flavor, Gas Chromatographic and Mass Spectral Identification of Neutral Components of Aroma Faction and Blue Cheese. *Journal of Agricultural and Food Chemistry.* 13(1):2-4.

p.48 BAKED BEANS

Steggerda FR. 1968. Gastrointestinal gas following food consumption. Annals of the New York Academy of Sciences 150(1):57-66.

Suarez FL, Springfield J, Levitt MD. 1998. Identification of gases responsible for the odour of human flatus and evaluation of a device purported to reduce this odour. *Gut.* 43(1):100-104.

p.53 CARROTS

Smith W, Mitchell P, Lazarus R. 2008. Carrots, carotene and seeing in the dark. *Clinical & Experimental Ophthalmology.* 27(3-4):200-203.

p.54 **BEETROOT**

Mitchell SC. 2001. Food Idiosyncrasies: Beetroot & Asparagus. *Drug Metabolism & Disposition*. 29(2):539-543.

p.57 **POTATOES**

Friedman M, McDonald GM, Filadelfi-Keszi M. 2010. Potato Glycoalkaloids: Chemistry, Analysis, Safety, and Plant Physiology. *Critical Reviews in Plant Sciences*. 16(1):55-132.

p.58 **AVOCADO**

Kahn V. 1975. Polyphenol oxidase activity and browning of three avocado varieties. *Journal of the Science of Food and Agriculture*. 26(9):1319-1324.

McEvily AJ, Iyengar R, Otwell WS. 1992. Inhibition of enzymatic browning in foods and beverages. *Critical Reviews in Food Science & Nutrition*. 32(3):253-273.

Bates RP. 1968. The retardation of enzymatic browning in avocado puree and guacamole. In *Proc. Fla. State Hort. Soc.* Vol. 81:230-5.

p.61 **FOOD COLOURINGS**

Bateman B, Warner JO, Hutchinson E, Dean T, Rowlandson P, Gant C, Stevenson J. 2004. The effects of a double blind, placebo controlled, artificial food colourings and benzoate preservative challenge on hyperactivity in a general population sample of preschool children. *Archives of Disease in Childhood*. 89(6):506-511.

Erickson B. 2011. Food dye debate resurfaces. *Chemical and Engineering News*, 27-31.

p.62 **SALMON**

Higuera-Ciapara I, Felix-Valenzuela L, Goycoolea FM. 2006. Astaxanthin: a review of its chemistry and applications. *Critical Reviews in Food Science and Nutrition*. 46(2):185-196.

p.65 **TONIC WATER**

Sacksteder L, Ballew RM, Brown EA, Demas JN. 1990. Photophysics in a disco: Luminescence quenching of quinine. *Journal of Chemical Education*. 67(12):1065.

p.69 **KIDNEY BEANS**

Rodhouse JC, Haugh CA, Roberts D, Gilbert RJ. 1990. Red kidney bean poisoning in the UK: an analysis of 50 suspected incidents between 1976 and 1989. *Epidemiology & Infection*. 105(3):485-491.

p.70 **MUSHROOMS**

Tu A, 1992. *Handbook of Natural Toxins, Vol. 7: Food Poisoning*. CRC Press. 207-229.

Holsen DS, Aarebrot S. 1997. Poisonous mushrooms, mushroom poisons and mushroom poisoning: a review. *Tidsskrift for den Norske Laegeforening: Tidsskrift for Praktisk Medicin*.117(23):3385-3388.

p.73 **APPLE**

Bolarinwa IF, Orfila C, Morgan MR. 2014. Amygdalin content of seeds, kernels and food products commercially available in the UK. *Food chemistry*. 152:133-139.

p.74 **SHELLFISH**

Yasumoto T, Murata M, Oshima Y, Sano M, Matsumoto GK, Clardy J. 1985. Diarrhetic shellfish toxins. *Tetrahedron*. 41(6):1019-1025.

Todd EC. 1993. Domoic acid and amnesic shellfish poisoning: a review. *Journal of Food Protection*. 56(1):69-83.

Watkins SM, Reich A, Fleming LE, Hammond R. 2008. Neurotoxic shellfish poisoning. *Marine Drugs*. 6(3):431-455.

Popkiss MEE, Horstman DA, Harpur D. 1979. Paralytic shellfish poisoning. *South African Medical Journal*. 55:1017-1023.

p.77 **PUFFERFISH**

Noguchi T, Hwang DF, Arakawa O, Sugita H, Deguchi Y, Shida Y, Hashimoto K. 1987. *Vibrio alginolyticus*, a tetrodotoxin-producing bacterium, in the intestines of the fish Fugu vermicularis vermicularis. *Marine Biology*. 94(4):625-630.

Ahasan HA, Mamun AA, Karim SR, Bakar MA, Gazi EA, Bala CS. 2004. Paralytic complications of pufferfish (tetrodotoxin) poisoning. *Singapore Medical Journal*. 45(2):73-74.

p.78 **CHOCOLATE**

Meng CC, Jalil AMM, Ismail A. 2009. Phenolic and theobromine contents of commercial dark, milk and white chocolates on the Malaysian market. *Molecules*. 14(1):200-209.

p.80 **HANGOVER**

Swift R, Davidson D. 1998. Alcohol hangover. Alcohol Health Res World. 22:54-60.

Rohsenow DJ, Howland J, Arnedt JT, Almeida AB, Minsky S, Kempler CS, Sales S. 2010. Intoxication With Bourbon Versus Vodka: Effects on Hangover, Sleep, and Next-Day Neurocognitive Performance in Young Adults. Alcoholism: Clinical and Experimental Research. 34(3):509-518.

p.85 **ONION**

Benkeblia N, Lanzotti V. 2007. Allium Thiosulfinates: Chemistry, Biological Properties and their Potential Utilization in Food Preservation. *Food*. 1(2):193-201.

Imai S, Tsuge N, Tomotake M, Nagatome Y, Sawada H, Nagata T, Kumagai H. 2002. Plant biochemistry: An onion enzyme that makes the eyes water. *Nature*. 419:685.

p.86 **CHILLI**

Bellringer M. The Chemistry of Chilli Peppers (online). Bristol: The University of Bristol. Available from: http://www.chm.bris.ac.uk/motm/chilli/index.htm (accessed 15.01.2014).

p.89 **MINT**

Hensel H, Zotterman Y. 1951. The effect of menthol on the thermoreceptors. *Acta physiologica Scandinavica*. 24(1):27-34.

p.90 **POPPING CANDY**

Sung AA, Lee YD. 1996. Gasified candy. *Trends in Food Science and Technology*. 7(6):205-205.

p.92 **WASABI**

Depree JA, Howard TM, Savage GP. 1998. Flavour and pharmaceutical properties of the volatile sulphur compounds of Wasabi (*Wasabia japonica*). Food research international. 31(5):329-337.

p.97 **TURKEY**

Lenard NR, Dunn AJ. 2005. Mechanisms and significance of the increased brain uptake of tryptophan. *Neurochemical Research*. 30(12):1543-1548.

p.98 **CHEESE**

Smith D. 2013. Sweet Dreams are Made of Cheese (online). Nature Publishing Group. Accessible at: http://www.nature.com/scitable/blog/mind-read/sweet_dreams_are_made_of.

p.101 **NUTMEG**

Shulgin AT, Sargent T, Naranjo C. 1967. The chemistry and psychopharmacology of nutmeg and of several related phenylisopropylamines. *Psychopharmacology bulletin*. 4(3):13-13.

Carstairs SD, Cantrell FL. 2011. The spice of life: an analysis of nutmeg exposures in California. *Clinical Toxicology*. 49(3):177-180.

p.102 **TEA**

Chin JM, Merves ML, Goldberger BA, Sampson-Cone A, Cone EJ. 2008. Caffeine content of brewed teas. *Journal of Analytical Toxicology*. 32(8):702-704.

Owen GN, Parnell H, De Bruin EA, Rycroft JA. 2008. The combined effects of L-theanine and caffeine on cognitive performance and mood. *Nutritional neuroscience*. 11(4):193-198.

p.105 **ABSINTHE**

Lachenmeier DW, Nathan-Maister D, Breaux TA, Luauté JP, Emmert J. 2010. Absinthe, Absinthism and Thujone–New Insight into the Spirit's Impact on Public Health. *Open Addiction Journal*. 3:32-38.

Padosch SA, Lachenmeier DW, Kröner LU. 2006. Absinthism: a fictitious 19th century syndrome with present impact. Substance abuse treatment, prevention, and policy. 1(1):14.

Lachenmeier DW, Nathan-Maister D, Breaux TA, Sohnius EM, Schoeberl K, Kuballa T. 2008. Chemical composition of vintage preban absinthe with special reference to thujone, fenchone, pinocamphone, methanol, copper, and antimony concentrations. *Journal of agricultural and food chemistry*. 56(9):3073-3081.

p.106 **ENERGY DRINKS**

Aranda M, Morlock G. 2006. Simultaneous determination of riboflavin, pyridoxine, nicotinamide, caffeine and taurine in energy drinks by planar chromatography-multiple detection with confirmation by electrospray ionization mass spectrometry. *Journal of Chromatography A*. 1131(1):253-260.

Higgins JP, Tuttle TD, Higgins CL. 2010. Energy beverages: content and safety. In Mayo Clinic Proceedings. 85(11):1033-1041. Elsevier.

p.111 **GRAPEFRUIT**

PL Detail Document. 2007. Potential Drug Interactions with Grapefruit. Pharmacist's Letter/Prescriber's Letter. 23(2):230204.

p.112 **LEMON**

Baron JH. 2009. Sailors' scurvy before and after James Lind–a reassessment. *Nutrition reviews*. 67(6):315-332.

p.115 **NUTS**

Bansal AS, Chee R, Nagendran V, Warner A, Hayman, G. 2007. Dangerous liaison: sexually transmitted allergic reaction to Brazil nuts. *Journal of Investigational Allergology and Clinical Immunology*. 17(3):189-191.

Fleischer DM. 2007. The natural history of peanut and tree nut allergy. Current allergy and asthma reports. 7(3):175-181.

p.116 **MEAT ALLERGY**

Saleh H, Embry S, Nauli A, Atyia S, Krishnaswamy G. 2012. Anaphylactic reactions to oligosaccharides in red meat: a syndrome in evolution. *Clin Mol Allergy*. 10(5).

p.119 **CLOVES**

Chaieb K, Hajlaoui H, Zmantar T, Kahla-Nakbi AB, Rouabhia M, Mahdouani K, Bakhrouf A. 2007. The chemical composition and biological activity of clove essential oil, *Eugenia caryophyllata* (*Syzigium aromaticum L. Myrtaceae*): a short review. *Phytotherapy research*. 21(6):501-506.

Kong X, Liu X, Li J, Yang Y. 2014. Advances in Pharmacological Research of Eugenol. *Curr Opin Complement Altemat Med*. 1(1):8-11.

p.120 **MONOSODIUM GLUTAMATE (MSG)**

Tarasoff L, Kelly MF. 1993. Monosodium L-glutamate: a double-blind study and review. *Food and chemical toxicology*. 31(12):1019-1035.

Williams AN, Woessner KM. 2009. Monosodium glutamate 'allergy': menace or myth?. *Clinical & Experimental Allergy*. 39(5):640-646.

Ng T. 2002. Re-evaluation of the Tasty Compound: MSG. *Nutrition Bytes*. 8(1).

p.123 **SWEETENERS**

Emsley J. 1994. *The consumers good chemical guide: a jargon-free guide to the chemicals of everyday life*. W.H. Freeman & Co. Ltd. 31-59.

O'Brien-Nabors L. (Ed.). 2011. *Alternative sweeteners* (Vol. 48). CRC Press.

Suez J, Korem T, Zeevi D, Zilberman-Schapira G, Thaiss C A, Maza O, Israeli D, Zmora N, Shlomit G, Weinberger A, Kuperman Y, Harmelin A, Kolodkin-Gal I, Shapiro H, Halpern Z, Eran S, Elinav E. 2014. Artificial sweeteners induce glucose intolerance by altering the gut microbiota. *Nature*. 514:181-186.

p.124 **SULFITES**

Taylor SL, Higley NA, Bush RK. 1986. Sulfites in foods: uses, analytical methods, residues, fate, exposure assessment, metabolism, toxicity, and hypersensitivity. *Advances in Food Research*. 30:1-76.

Lund B, Baird-Parker TC, Gould GW. 2000. Microbiological safety and quality of food (Vol. 1). Springer. 201-203.

p.129 **BANANA**

Burg SP, Burg EA. 1965. Relationship between ethylene production and ripening in bananas. *Botanical Gazette*. 200-204.

Jiang Y, Joyce DC, Macnish AJ. 1999. Extension of the shelf life of banana fruit by 1-methylcyclopropene in combination with polyethylene bags. *Postharvest Biology and Technology*. 16(2):187-193.

p.130 **JELLY**

Jacobsen E. 1999. Soup or Salad? Investigating the Action of Enzymes in Fruit on Gelatin. *Journal of Chemical Education*. 76(5):624A.

p.133 **WHIPPED CREAM**

Noda M, Shiinoki Y. 1986. Microstructure and rheological behavior of whipping cream. *Journal of Texture Studies*. 17(2):189-204.

Dickinson E, Stainsby G. 1982. Colloids in food. Applied Science Publishers.

p.134 **CHOCOLATE**

Langer S, Marshall LJ, Day AJ, Morgan MR. 2011. Flavanols and methylxanthines in commercially available dark chocolate: a study of the correlation with nonfat cocoa solids. *Journal of Agricultural and Food Chemistry*. 59(15):8435-8441.

p.137 **BEER**

Vogler A, Kunkely H. 1982. Photochemistry and beer. *Journal of Chemical Education*. 59(1):25.

p.138 **JAM**

Thakur BR, Singh RK, Handa AK. 1997. Chemistry & Uses of Pectin – A Review. *Critical Reviews in Food Scienc & Nutrition*. 37(1):47-73.

p.140 **RED WINE**

Alcalde-Eon C, Escribano-Bailón MT, Santos-Buelga C, Rivas-Gonzalo JC. 2006. Changes in the detailed pigment composition of red wine during maturity and ageing: a comprehensive study. *Analytica Chimica Acta*. 563(1):238-254.

Boulton R. 2001. The copigmentation of anthocyanins and its role in the color of red wine: a critical review. *American Journal of Enology and Viticulture*. 52(2):67-87.

Cheynier V, Dueñas-Paton M, Salas E, Maury C, Souquet JM, Sarni-Manchado P, Fulcrand H. 2006. Structure and properties of wine pigments and tannins. *American Journal of Enology and Viticulture*. 57(3):298-305.

Semba RD, Ferrucci L, Bartali B, Urpí-Sarda M, Zamora-Ros R, Sun K, Andres-Lacueva C. 2014. Resveratrol Levels and All-Cause Mortality in Older Community-Dwelling Adults. JAMA internal medicine.

p.143 **CHAMPAGNE**

Liger-Belair G, Cilindre C, Gougeon RD, Lucio M, Gebefügi I, Jeandet P, Schmitt-Kopplin P. 2009. Unraveling different chemical fingerprints between a champagne wine and its aerosols. Proceedings of the National Academy of Sciences. 106(39):16545-16549.

Liger-Belair G. 2005. The physics and chemistry behind the bubbling properties of Champagne and sparkling wines: A state-of-the-art review. *Journal of Agricultural and Food Chemistry*. 53(8): 2788-2802.

ACKNOWLEDGEMENTS

Primarily, I'd like to thank Emma Smith and the team at Orion who've helped to make this book a reality, and worked hard to hone its content. Thanks go, too, to the numerous people (too many to list individually here) who have helped with providing studies and discussions relating to the chemistry covered in this book, particularly via Twitter.

Additional thanks goes to Professor Matthew Hartings of American University, Washington DC, for his assistance proof-reading content for the book.

The chemical structures in the graphics are created using PerkinElmer's ChemDraw Professional v15, which is available here: http://bit.ly/1lhoZ8g. Compound Interest is independent of and has no association with PerkinElmer or its affiliates.

SOCIAL MEDIA

@compoundchem

tumblr.com/blog/compoundchem

facebook.com/compoundchem

PROGRAMS & SITES

- Chemical structures – ChemDraw
- Graphic design – InDesign
- Selected CC0 icons – The Noun Project
- Fonts: Open Sans, Bebas Neue, Lobster, Varela Round, Montserrat, Helvetica Neue, Oswald

SEE MORE FROM COMPOUND INTEREST
WWW.COMPOUNDCHEM.COM

Visit for a wide range of graphics on various aspects of everyday chemistry, as well as free graphic downloads, purchasable posters, and more.

This edition first published in Great Britain in 2015 by
Orion
an imprint of the Orion Publishing Group Ltd
Carmelite House,
50 Victoria Embankment,
London EC4Y 0DZ

An Hachette UK Company

10 9 8 7 6 5 4 3 2 1

A CIP catalogue record for this book is available from the British Library.

ISBN: 978 1 4091 5661 1

Design Concept by Grade Design

Printed in Italy

www.orionbooks.co.uk

Picture credits:

Andy Brunning: page 21, 25, 26, 29, 30, 37, 38, 52, 56, 60, 72, 79, 84, 87, 88, 91, 100, 103 (left), 110, 112, 118, 128, 143; Food and Drink/REX Shutterstock: page 131; Getty: page 18; Shutterstock.com: page 17, 22, 33, 41, 42, 45, 46, 49, 55, 59, 63, 68, 76, 93, 99, 103 (right), 107, 114, 117, 136, 140; REX Shutterstock: page 139.